Xue Liu

Instability and Segregation in bounded Gas-Particle and Granular Flows

Xue Liu

Instability and Segregation in bounded Gas-Particle and Granular Flows

VDM Verlag Dr. Müller

Imprint

Bibliographic information by the German National Library: The German National Library lists this publication at the German National Bibliography; detailed bibliographic information is available on the Internet at
http://dnb.d-nb.de.

Cover image: www.purestockx.com

Published 2008 Saarbrücken

Publisher:
VDM Verlag Dr. Müller Aktiengesellschaft & Co. KG, Dudweiler Landstr. 125 a,
66123 Saarbrücken, Germany,
Phone +49 681 9100-698, Fax +49 681 9100-988,
Email: info@vdm-verlag.de
Zugl.: Piscataway NJ, Rutgers University, Dissertation, 2006

Produced in Germany by:
Schaltungsdienst Lange o.H.G., Zehrensdorfer Str. 11, 12277 Berlin, Germany
Books on Demand GmbH, Gutenbergring 53, 22848 Norderstedt, Germany

Impressum

Bibliografische Information der Deutschen Nationalbibliothek: Die Deutsche Nationalbibliothek verzeichnet diese Publikation in der Deutschen Nationalbibliografie; detaillierte bibliografische Daten sind im Internet über http://dnb.d-nb.de abrufbar.

Coverbild: www.purestockx.com

Erscheinungsjahr: 2008
Erscheinungsort: Saarbrücken

Verlag: VDM Verlag Dr. Müller Aktiengesellschaft & Co. KG, Dudweiler Landstr. 125 a,
D- 66123 Saarbrücken,
Telefon +49 681 9100-698, Telefax +49 681 9100-988,
Email: info@vdm-verlag.de
Zugl.: Piscataway NJ, Rutgers University, Dissertation, 2006

Herstellung in Deutschland:
Schaltungsdienst Lange o.H.G., Zehrensdorfer Str. 11, D-12277 Berlin
Books on Demand GmbH, Gutenbergring 53, D-22848 Norderstedt

ISBN: 978-3-8364-8975-1

Abstract of the Dissertation

Instability and Segregation in Bounded Gas-Particle Fluidized Beds

and Granular Flows

By Xue Liu

Dissertation Director:
Professor Benjamin J. Glasser

The objective of this work is to advance our understanding of granular and gas-particle flows, using continuum models based on kinetic theory. In gas-particle fluidized beds, confined between parallel solid walls, non-uniform solids distribution is observed with a mono-disperse particle mixture. Using a bifurcation/continuation approach, it is found that the steady state flow profiles are strongly related to the physical and operating parameters, such as particle inelasticity, gravity, bed width and mean solids fraction. A linear stability analysis of equations of motion describing gas-particle fluidized beds has been carried out to investigate conditions at which the gas-particle flows become unstable. It is shown that the dominant instability pattern has two types of structures: symmetric ones and anti-symmetric ones. Both of them propagate through the bed in the form of a traveling wave at a speed comparable with that of the solids phase. The symmetric instabilities can develop into bubble-like structures while the anti-symmetric instabilities can develop into streamer-like structures. All instabilities reported in this work are suppressed if solid phase inertia or inelastic particle collisions are removed.

Finally, we extend our work to investigate the steady state solutions of boundary-driven and body-force-driven granular flows and gas-particle fluidized beds with a bi-disperse particle mixture. Granular energy equipartition breaks down with an increase in the system inelasticity and the mass ratio, whereas the effect of the size ratio is very small if the two particle species have the same mass. The segregation of total particles and each individual particle species is enhanced when the effect of non-equipartition of granular energy is considered. The species segregation is enhanced with a decrease in the system elasticity, an increase in the average solids fraction or an increase in the size ratio, due to the competition of three diffusion forces: the thermal diffusion force, the ordinary diffusion force and the pressure diffusion force. For gas-particle fluidized beds and granular flows in a channel, we find that for the kinetic theory models used in the present study, the solids fraction and the granular temperature profiles are quite similar.

Acknowledgments

Foremost, I would like to thank my research advisor, Prof. Benjamin J. Glasser for guidance, encouragement and support during my time at Rutgers. This research has benefited immensely from his seemingly endless access to interesting ideas, unwavering enthusiasm and a healthy dose of skepticism. I thank my committee members, Profs. Marianthi Ierapetritou, Alberto Cuitino as well as Dr. Stelios C. Tsinontides for their time and inputs on my thesis work.

This work was supported by the National Science Foundation (NSF), and National Aeronautics and Space Administration (NASA).

Many thanks for the support of my fellow graduate researchers, including Stephen, Jodi, Azzeddine, Jayati, Kiernan, Athanas, Justin, Marcus, Mobeen. Thanks extend to my friends and my family, my Mum and Dad. I finally thank my wife, Dan Wang, for her brilliant support on my work and my life.

Table of Contents

List of Tables

List of Figures

Chapter 1 Introduction

1.1 Overview

Fluidized bed reactors consist of a bed of solids that comprise the particulate phase, through which a gas is passed. When the gas flow rate through the bed is increased, the pressure drop across the bed increases linearly until its magnitude equals to the weight of the bed. At this critical flow velocity the gas-particle system behaves like a fluid, and is said to be "fluidized" (Wilhelm and Kwauk, 1948).

Generally, gas and liquid fluidized beds behave quite differently. In liquid-solid systems, an increase in flow rate beyond the minimum fluidization usually results in a smooth expansion of the bed, which is called a homogeneously fluidized bed. In most gas-solid systems, however, with an increase in gas flow rate above the minimum fluidization, instabilities with bubbling and channeling of gas are observed. Normally, the classification of gas-fluidized beds is based on the gas flow rate. When the gas velocity is beyond the minimum fluidization, regions containing only gas form and grow as they rise through the bed. These voids burst through the upper surface of the bed, giving the bed an appearance of a boiling liquid. This is called a bubbling fluidized bed. In a long and narrow system, bubbles can span the width of the column and particles above bubbles are pushed upward as by a piston. This is described as a slug. When particles are fluidized by a sufficiently high gas flow rate, the upper surface of the bed disappears, and instead of bubbles, a turbulent motion of solid clusters and voids of gas with different sizes and shapes can be observed. This is the turbulent fluidized bed. With a further increase in gas flow rate, particles are carried out of the bed with the gas and a cyclone is required for steady state operations. This system is called a fast fluidized bed with pneumatic transport of particles (Geldart, 1973, 1986; Jackson, 2000; Kunii and Levenspiel, 1991).

The use of fluidized beds can be dated to the development of coal combustion in Germany in the 1920s, and fluidization technology received tremendous attention during the development of fluidized catalytic cracking of petroleum feedstock (Kunii and Levenspiel, 1991). At the present time, gas-particle fluidized bed reactors are used extensively in continuous petroleum, chemical, pharmaceutical, agricultural, biochemical, food, electronic, and combustion processes. A diversity of designs of fluidized beds has been developed for a tremendous variety of applications such as drying, tablet coating, catalytic cracking and the cultivation of microorganisms (Fan and Chao, 1998).

Fluidized beds are used widely in industry primarily because they posses very high rates of heat and mass transfer due to excellent mixing between the gas and the solid phases (Kunii and Levenspiel, 1991). The smooth flow of particles allows continuously controlled operations with easy handling. The rapid mixing of solids resists rapid temperature changes, leading to an isothermal condition throughout the reactor. Therefore, a large margin of safety in avoiding temperature runaways for highly exothermic reactors can be obtained using fluidized bed reactors. Using cyclones to collect particles or using multiple fluidized beds to accomplish circulation makes it possible to remove or add vast quantities of heat in reactors, making large-scale operations possible.

In some gas-particle fluidized beds, it is possible to expand the bed homogeneously or uniformly (Geldart, 1986; Tsinontides and Jackson, 1993). However, from the earliest application of gas-particle fluidized beds, it has been recognized that the uniformly fluidized bed is only one possible flow regime. In most gas-particle fast fluidized beds, the particles are seen normally to segregate near the walls and the core of the reactor is occupied by a dilute region with high gas velocities (Bader et al., 1988; Nieuwland et al., 1996; Weinstein et al., 1984; Yerushalmi et al., 1978). Clearly then, the flow characteristics, such as pressure gradient and velocity profile, are not the same as if the particles are uniformly distributed. Therefore, it is important to understand the physical mechanism responsible for this heterogeneity and to be able to predict this behavior quantitatively. In addition, when particles are propelled rapidly upward by gas in a fluidized bed, voidage waves, such as bubbles, slugs, clusters and streamers are observed to develop and move through the bed. The unsteady gas-particle flow and associated fluctuations in the particle concentration and pressure drop can lead to poor yield and selectivity of chemical reactions in fluidized beds, resulting in excessive production of byproducts which require disposal or downstream processing. The occurrence of such instability waves can have a significant effect on the economics, safety, and environmental impacts of fluidized bed reactors. The variety of possible flow regimes in gas-fluidized beds can lead to difficulties with the scale-up because the gas bypassing and gas-particle mixing rate, as well as other pertinent quantities, all vary as the flow regime changes (Jayati and Glasser, 2002). At the present time, fluidized beds are designed based on correlations and empiricism, and costly pilot plants must be built for each and every process because there is very little understanding of which flow regimes can be observed for a given set of operation conditions, and the physical mechanism leading to the occurrence of the voidage or density waves is still not completely understood (Davidson et al., 1985; Geldart, 1986; Homsy, 1998; Jackson, 1971, 2000).

The goal of this research is to improve our understanding of the hydrodynamics in granular flows and gas-particle fluidized beds. A continuum model based on kinetic theory is developed to describe gas-particle flows in a channel. The effects of physical and operational parameters, and dimensionless terms such as the Reynolds number and the Froude number, on the steady state flow profiles are examined using a continuation approach. In order to understand the stability behavior of the base state solutions, the governing equations and boundary conditions are analyzed using a linear stability analysis. It is found that the dominant instabilities have the form of traveling waves. The density wave structures may be symmetric or anti-symmetric about the centerline of the channel. To better understand the effect of the flow hydrodynamics on the chemical reactions in gas-particle fluidized beds, a simple chain reaction is assumed, and the time dependent evolution of the component concentration profiles are examined. Finally, a model involving a binary particle mixture is considered and the solutions of steady fully developed flows are computed for bounded granular flows (Couette flow and channel flow), and gas-particle fluidized beds. The effects of physical parameters and operating conditions on the flow profiles and segregation behaviors are examined.

1.2 Equations of Motions

There are two approaches commonly used to model the flow behavior of particles flowing with a carrier gas: the Eulerian approach (Dasgupta et al., 1994, 1998; Detamore et al., 2000; Hrenya and Sinclair, 1997; Louge et al., 1991; Ocone et al., 1993; Pita and

Sundaresan, 1991; Sinclair, 1989; Yasuna et al., 1995; Zhang and Reese, 2001) and the mixed Eulerian-Lagrangian approach (Goldschmidt et al., 2001; Helland et al., 2000; Li and Kuipers, 2003; Zhou et al., 2002). Eulerian models, or two fluid models, treat both the gas and solid particles as a continuum, based on volume or ensemble averaged equations of motion. Mixed Eulerian-Largrangian models, or discrete particle models, treat the gas phase as a continuum while computing the trajectory of each particle. The interactions between the gas phase and the particles are described based on experimental and theoretical correlations. The advantage of this method is that the particle-particle and particle-wall interactions can be derived using simple physical laws (Tsuji et al., 1993; Xu and Yu, 1997). Thus, the uncertainties involved in the closures for the solids phase, such as particle pressure and solids phase viscosity are eliminated. Therefore, discrete particle simulation is a powerful tool used to examine constitutive relations derived in continuum theories. The difficulty in the application of this approach is that an immense computational resource is required to follow the movement of each particle and calculate each contact or collision between particles or between particles and the walls. From a computational point of view, the Eulerian approach is attractive because the calculations are not limited by the number of particles in the system, which is advantageous when modeling large scale reactors.

Equations with averaged forms have been extensively proposed in the literature (Anderson and Jackson, 1967; Drew, 1971; Jackson, 1997; Wallis, 1991). Some authors used local spatial averages taken over regions large enough to contain many particles but small compared to the macroscopically interested length scale; others used ensemble averages that average at each point of space over an ensemble of "macroscopically equivalent" systems. The averaging process generates averaged quantities, such as solid phase pressure and solid phase viscosity, which require closure through constitutive relations. This has been proved to be difficult and has been achieved only for some simple cases (Jackson, 2000).

It is common to derive the constitutive relations for particles experiencing rapid deformation from the kinetic theory of dense gases (Jenkins and Richman, 1985; Jenkins and Savage, 1983; Lun, 1991; Lun et al., 1984). The main difference between the molecular gas and the granular particles is that energy is lost from collisions between particles, while no energy is lost from molecular gas collisions. In the kinetic theory of granular materials, the particle phase has both an effective pressure and an effective viscosity, entirely analogous to the pressure and viscosity of a molecular gas. For a gas, these variables depend on the thermal temperature. Analogously, for a granular material, the particle pressure and particle viscosity are determined by the granular temperature, which is proportional to the mean square of particle fluctuation velocity. Therefore, a supplementary equation to describe the balance of pseudo-thermal energy is required to close the mass and momentum balance equations.

Two kinetic theory models have been developed to model systems containing a large number of molecules, the Vlosoc (Bellomo and Pulvirenti, 2000) model, and the Boltzmann model. In the Vlosoc model, the external force applied on each molecule is derived from a potential that depends on the locations and motions of other molecules in the system. Basically, this model is suitable to plasma gases. The Boltzmann model is an evolution equation based on the velocity distribution function of molecules (Jenkins and Richman, 1985; Jenkins and Savage, 1983; Lun, 1991; Lun et al., 1984). Using this model, the thermodynamic properties can be explained from purely mechanical systems. Thus, this model is suitable to describe the motions of granular flows or fluid-particle flows.

Chapman and Cowling (1970) developed a kinetic theory to predict the behavior of mass point molecules. In their model the interaction energies between particles are assumed conserved. Ogawa et al. (1980) suggested that the dissipation of mechanical energy is different in granular flows. Generally, the mechanical energy can be directly transformed into internal energy based on conventional dissipation mechanisms. In granular systems, however, the mechanical energy is first transformed into random particle motion and then dissipated into internal energy. To consider the energy dissipation arising from particle-particle collisions, Savage and Jeffery (1981) extended kinetic theory to rapidly deforming smooth hard spherical particles whose interaction energies are not conserved. They modified the classical treatment of the collision between two particles by introducing the concept of the coefficient of restitution. They assumed the single-pair velocity distribution function was Maxwellian and calculated the particle stress tensor. The weakness of their treatment is its restriction to a mean simple shear flow. This constraint was removed by Jenkins and Savage (1983), whose model can be applied to any type of flow. In their model (Jenkins and Savage, 1983) the value of the coefficient of restitution was assumed to be close to unity, which is not the case for many industrially relevant particles.

In the past two decades, a number of investigators contributed to the development of the kinetic theory approach (Ding and Gidaspow, 1990; Goldshtein and Shapiro, 1995; Johnson and Jackson, 1987; Lun, 1991; Lun and Savage, 1987; Lun et al., 1984;

Montanero et al., 1999; Sela and Goldhirsh, 1998). Lun et al. (1984) presented two kinetic theories for the rapid flows of cohesionless granular materials. The first one followed the work of Savage and Jeffery (1981), and used an *ad hoc* distribution function to describe the collisions between particles. The second one determined the distribution function as part of the overall analysis using Chapman-Enskog theory, but was intended for nearly elastic particles. Therefore, the second approach could predict the collisional distribution functions by a rational approximation scheme, and incorporated kinetic as well as collisional contributions to the constitutive relations for particle stress and energy flux. The prediction results from the second approach have been used in a large number of papers to simulate granular flows (Alam and Nott, 1998; Nott et al., 1999; Wang et al., 1996, 1997; Wang and Tong, 1998, 2001) or gas-particle flows (Agrawal et al., 2001; Hrenya and Sinclair, 1997; Ocone et al., 1993; Pita and Sundaresan, 1991; Sinclair and Jackson, 1989). Lun and Savage (1987) added the rotation and surface frictional forces into the kinetic theory of granular materials and showed that this model had better agreement with experimental results for dense systems because frictional forces may dominate dense granular flows where sustained contacts can be expected to be more important. Lun et al. (1991) extended this work by considering the contribution of spin and sliding friction in the calculation. Thus, their model is appropriate for both dilute and dense systems. Sela et al. (1996) studied simply sheared two-dimensional granular disks in a plane, and showed that the normal stress differences observed in granular flows is due to the Burnett correction to the single particle distribution function. In a following paper, Sela and Goldhirsch (1998) derived hydrodynamic equations for rapid flows of smooth inelastic spheres in three dimensions. The constitutive relations shown in their paper differ from those obtained in previous studies. This is due to the fact that the Sela and Goldhirsch model defines a relevant short time scale for the hydrodynamics of the granular temperature field. Ding and Didaspow (1990) extended kinetic theory to model two-phase flows. They developed governing equations as well as some bulk phase granular properties for gas-particle flows based solely on first principles. The boundary conditions for granular flows adjacent to solids walls were investigated by Johnson and Jackson (1987; 1990).

So far the majority of in-depth research on granular materials have been confined to mono-disperse systems, where the particles have equal density and equal size. However, it is well known that in practice the particle size distribution plays an important role in the behavior of granular or gas-particle mixtures. Segregation by particle size and density is a common behavior in many processes, such as shearing, shaking and pouring (Bagnold, 1954; Bridgewater, 1976, 2003; Bridgewater et al., 1978; Bridgewater et al., 1985; Savage and Sayed, 1984). Size segregation can occur when small particles occupy the gaps between large particles, while density segregation occurs when heavier particles move away from lighter particles and segregate together.

Theoretical models and experiments have been established to study the segregation behaviors observed in polydisperse particle mixtures. Bagnold (1954) observed that in particle mixtures large grains have more potential to migrate towards the regions where the shear strain is low, such as the free surface in a gravity channel flow. Arastoopour et al. (1982) developed a model to simulate a one dimensional flow in a vertical pneumatic conveying line with a mixture of a gas and a polydisperse solid phase. They showed that the particle size distribution affects the extent of the particle segregation and the variation of the pressure drop along the vertical transport line. Experimental works also showed that the flow of a poly-disperse particle mixture does not have the same behavior as a mono-disperse particle mixture. Savage and Sayed (1984) did experiments on shear flows of a binary particle mixture with the same density but different sizes. They showed that the stresses for a binary mixture was five times lower than those for a mono-size mixture. Many investigations of segregation phenomena during particulate material shear deformation were carried out by Bridgewater and coworkers (Bridgewater, 1976; Bridgewater et al., 1978; Bridgewater et al., 1985) to study the mechanics of percolation and segregation of dry granular materials. Particle size segregation in inclined chute flows of dry cohesionless granular solids was investigated experimentally and theoretically by Savage and Lun (1988), who used a percolation mechanism to explain the development of the solids concentration profiles with downstream distance. Dolgunin et al. (1998) studied the segregation of solid spherical particles differing both in size and in density during a rapid gravity flow down a rough inclined chute, and showed that the migration of particles is caused by the solids concentration and the granular energy gradients. Arnarson and Willits (1998) computed thermal diffusion factors to characterize the segregation of granular materials due to granular energy gradients, based on the work of Kincaid et al. (1987). The thermal diffusion factor gives the magnitude and direction of the segregation relative to the energy gradient. Khakhar et al. (1999) derived expressions for the segregation flux due to pressure gradients and temperature gradients in a granular chute flow. They found when particles are elastic or nearly elastic, theoretical predictions are in good agreement with particle dynamic simulations or Monte Carlo simulations if the average solids fraction is low or moderate (<0.2). The effect of density segregation (equal size particles with different density) and size segregation (equal density particles with different size) was also analyzed in this paper. The theory showed that density segregation drives denser particles to sink to lower regions, while size segregation results in a net downward flux of larger particles.

To better understand the behavior of granular flows or fluid-particle flows with poly-disperse particle mixtures, constitutive models for poly-disperse systems have been reported since the early 1990's, following the analogy with the kinetic theory of dense gas mixtures (Alam and Luding, 2003; Arnarson and Willits, 1998; Jenkins and Mancini, 1987, 1989; Lun and Savage, 1986; Willits and Arnarson, 1999). Jenkins and Mancini (1987) extended the kinetic theory of a dense gas to a binary mixture of smooth, slightly inelastic hard disks. They assumed the deviation of the component granular energy from the mixture granular energy was very small. Therefore, their model is only suitable for a mixture which can be characterized by the assumption of energy equipartition. Another assumption of their model was that the single velocity distribution had the form of a Maxwellian distribution function. This assumption results in an under-prediction of the granular energy flux and the granular stress because the contribution of the kinetic pressure tensor is under-predicted and the value of the kinetic heat flux is equal to zero (Willits and Arnarson, 1999). Arnarson and Willits (1998) and Willits and Arnarson (1999) followed the work of Jenkins and Mancini (1989), using the Chapman and Enskog procedure to solve the Boltzmann equation. In the procedure, the Maxwellian velocity distribution is expanded by adding a perturbation, which is a function of particle properties and operational parameters. Their models were restricted to energy equiparition. Theoretical studies have been complemented by both computer simulations (Alam and Luding, 2002b, 2003; Garzo and Dufty, 1999; Huilin et al., 2001; Huilin et al., 2000; Huilin et al., 2003) and experimental measurements (Feitosa and Menon, 2002; Wildman and Parker, 2002), verifying the breakdown of the equipartition principle for granular mixtures. As far as we know, Garzo and Dufty (1999) were the first to show that energy equipartition was violated for inelastic collisions of particles with different mechanical properties, using kinetic theory. Their results were experimentally confirmed by Wildman and Parker (2002) and Feitosa and Menon (2002), whose experiments showed that a binary granular mixture had two different granular temperatures when the inelastic particles were exposed to external vibrations. Although Garzo and Dufty (1999) solved the kinetic equations for systems away from equilibrium, their model was restricted to dilute systems, where the radial distribution function was close to 1. This constraint was lifted by Hadjira and Arastoopour (2005), who solved the Boltzman equation for each particulate phase using the Chapman-Enskog procedure by adding a perturbation to the Maxwallian velocity distribution function. They assumed at the collisional interface the perturbation could be transferred to the particulate bulk phase and hence at the interface the velocity distribution function remained Maxwellian. Using this model, the simulation of the rheological behavior of a binary mixture showed a good agreement with molecular dynamics simulations (Galvin et al. (2005) and experimental measurements (Feitosa and Menon (2002).

1.3 Fully-Developed Gas-Particle Flow in a Channel

A number of theoretical studies have been performed in the last fifteen years, to analyze steady fully developed gas-particle flows in a vertical tube, using kinetic theory. To the best of our knowledge, Sinclair and Jackson (1989) were the first to set up a laminar gas-laminar particle model to simulate fully developed gas-particle flows in a vertical pipe. By plotting the variation of the gas flowrate as a function of the particle flowrate for a given pressure drop, a contour map of the possible flow regimes was obtained. The performance diagrams were calculated over the whole range of cocurrent upflows and downflows, and countercurrent flows. The contour map and flow profiles qualitatively captured some important features of vertical pipe flows, such as packed beds, fluidized beds, choking and flooding. Yasuna et al. (1995) compared a large body of experimental measurements with the predictions from the models developed by Sinclair and Jackson (1989). In the paper of Yasuna et al. (1995), three types of drag closures were evaluated, and it was concluded that the flow behaviors are insensitive to the different drag force closures.

Particle inelasticity is a critical parameter for the stability of gas-particle systems. In the Sinclair and Jackson model (1989), two coefficients are used to describe the system elasticity. One is the coefficient of particle-particle restitution, e_p, the other is the coefficient of particle-wall restitution, e_w. The value of these two coefficients ranges from zero for perfectly inelastic collisions to one for perfectly elastic collisions. A main topic of discussion on the application of Sinclair and Jackson model is that the steady state profiles manifest an extreme sensitivity to the coefficient of particle-particle restitution

(Ocone et al., 1993; Pita and Sundaresan, 1991; Sinclair and Jackson, 1989). To eliminate this sensitivity, turbulent models have been proposed in several papers (Dasgupta et al., 1994, 1998; Eduardo et al., 1995; Hrenya and Sinclair, 1997; Louge et al., 1991). Dasgupta et al. (1994; 1998) removed the sensitivity of the steady state flow profiles to the coefficient of restitution by adopting a simple form for the closure of particle phase stress. Therefore, the particle phase stress is not dependent on the granular temperature, and the resulting equations have no reference to the coefficient of particle-particle restitution. Hrenya and Sinclair (1997) described two models to investigate the combined effects of the interactions among individual particles and the interactions associated with the collective motions of particles. By adding particle turbulence contributions into the governing equations and the constitutive relation for the dissipation of pseudo thermal energy, they found that the sensitivity of the system to inelastic particle-particle collisions could be reduced greatly. However, whether the sensitivity to the coefficient of restitution in the laminar model is really unrealistic is still not fully resolved. Previous work has indicated that in rapid granular flows, in which particles interact via nearly instantaneous collisions, the value of granular temperature is related to the magnitude of $1\text{-}e_p^2$ (Goldhirsch and Zanetti, 1993; Tan and Goldhirsch, 1997, 1998). For example, in a simple sheared system, when the effect of gravity is neglected, the granular temperature in the steady state flow converges to a value $T = C\frac{\gamma^2 l_0^2}{\varepsilon}$, where $\varepsilon = 1 - e_p^2$ is the degree of inelasticity, l_0 is the equilibrium mean free path, γ is the shear rate, and C is a volume fraction dependent prefactor (Goldhirsch, 1999). In this case, it is not surprising that the granular temperature is very sensitive to the coefficient of restitution when its value is close to 1, and this sensitivity is reduced as the value of the coefficient of restitution is decreased.

Wang et al. (1996; 1997) reported that for a granular flow in a channel, walls could be a source or a sink of fluctuational energy, depending on the value of e_w and e_p, where e_w describes the degree of inelasticity of collisions between particles and the walls. If e_w is smaller than e_p, the wall is a sink of fluctuational energy, and the granular temperature has a negative gradient towards the walls. When e_w is greater than e_p, the wall is a source of fluctuational energy, and the pseudo-thermal energy has a flux from the walls to the particle assembly. It is of interest to extend the study of the nature of boundaries to the gas-particle system to see their effects on gas-particle flow profiles.

Another parameter examined in this work is the gravity force, which has a significant effect on the fluidization state. Many operating parameters, such as the particle terminal velocity and the pressure gradient, vary with different values of the gravity force. An understanding of the effects of gravity on fluidized beds is important for local resource utilization on the Moon and Mars. Future missions to the Moon and Mars would likely make use of oxides in the soil and a fluidized bed is a likely candidate for processing lunar and Martian soil. Gravity is also an important parameter for the stability of fluidized suspensions. Based on the work of Gibilaro and Foscolo (1986), a significant stabilization of the fluidization process could be predicted under low gravity conditions. Compared with liquid-fluidized beds, this stabilization effect is more pronounced in gas-fluidized systems. However, the contributions of the gravity on the fully-developed flow solutions of gas-particle fluidized beds have not been widely investigated.

So far the majority of theoretical studies have been confined to narrow systems. In industrial systems the width of the units is usually many orders of magnitude larger than the particle diameter. Pita and Sundaresan (1991) used the Sinclair and Jackson model to study the scale-up characteristics of pipe flows for a large range of tube diameters. The occurrence of steady-state multiplicity wherein different pressure drops could be obtained for the same gas and particle flow rates was observed. They also found that for small tube diameter the pressure drop decreases rapidly as the bed diameter increases. However, when the tube diameter is sufficiently large, this trend can be reversed.

The contribution of the mean solids fraction on the stability of gas-fluidized beds has been studied in a large number of papers (Anderson and Jackson, 1968; Hirayama and Tsuzuki, 1995; Jackson, 1963). A stability analysis shows that an increase in solids fraction can suppress the instability because the particle pressure and the solid phase effective viscosity increase with increasing solids fraction (Anderson et al., 1995; Glasser et al., 1996, 1997). The variations of the steady state flow profiles with different average solids fractions were presented by Sinclair and Jackson (1989) for a narrow gas-fluidized bed. It is of interest to extend their analysis to wider systems to see the influence of the solid fraction on the steady state solutions.

In many papers, the governing equations were cast in dimensionless form (Ocone et al., 1993; Pita and Sundaresan, 1991; Sinclair and Jackson, 1989). Thus, dimensionless groups are introduced in the equations. In the present work, two dimensionless groups have been studied. One is the Reynolds number, Re, and the other is the Froude number, Fr. The Reynolds number can be an important parameter for the transition of flow regimes in gas-fluidized beds. Fr, which represents the ratio of inertia force to the gravity force, is also used to characterize fluidized beds.

1.4 Stability Analysis of Fluidized Beds

For unbounded gas-particle flows, the governing equations have a simple solution corresponding to the well-known state of uniform fluidization. To predict the growth of disturbances from this uniform state, the conventional hydrodynamic stability analysis has been applied (Anderson and Jackson, 1967, 1968; Pigford and Baron, 1965). The earliest stability analysis (Jackson, 1963) predicted that uniform fluidized beds should be always unstable and small disturbances could grow into voidage waves. However, the growth rate does not have a limit in the Jackson model (1963). Its value increases monotonically with a decrease in the perturbation wavelength. This physically unacceptable prediction is because the particle phase momentum equation does not contain a term to represent the effect of viscous dissipation. When this term was included in the model (Anderson and Jackson, 1967, 1968; Pigford and Baron, 1965), there is a critical wavelength, below which the system is stable. The results of linear stability analysis showed that the dominant disturbance has a form of a plane wave propagating vertically upward through the bed, and the growth rate for disturbances is larger by one or two orders of magnitude in typical gas fluidized beds than in typical liquid fluidized beds. Batchelor and Nitsche (1991) and Batchelor (1993) studied the linear stability of a stationary fluid with sinusoidal distribution of (particle) density in the vertical direction. They found that an instability occurs in a certain range of the lateral wave numbers and vertical wave numbers. Hirayama and Tsuzuki (1995) constructed a two-dimensional fluidized bed and investigated its linear stability to two-dimensional disturbances. Their results showed that the fluid inertia term is responsible for the instability, while the particle diffusion term and the particle viscosity term play major roles to suppress the growth of disturbances. Koch and Sangani (1999) studied the marginal stability limits for the homogeneous state of a gas-particle fluidized bed, using linear stability analysis. In their model, they did not use experiential correlations to describe the averaged terms in the equations of motion. Instead, all terms were derived from first principles.

From the earliest application of fluidized beds, it was noticed that most liquid fluidized beds have a smooth appearance when the beds are expanded, while most gas fluidized beds are heterogeneous containing bubbles, slugs or clusters. One of the earliest attempts to describe the mechanical behavior of different flow regimes came from Wilhelm and Kwauk (1948). They used the Froude number, which represents the ratio of inertia force to the gravity force, to characterize fluidized beds: $Fr = u_m^2 / d_p g$, where u_m denotes the minimum fluidization velocity and d_p is the particle diameter. If Fr is smaller than 1, the fluidized beds behave homogeneously, otherwise heterogeneously. This simple criterion is still widely quoted. To differentiate the bubbling and non-bubbling behavior in the fluidized beds, the technique of linear stability analysis has been used in many studies. Although linear theory can account for the existence of slowly growing waves in liquid-fluidized beds and much faster growing waves in gas-fluidized beds, it cannot explain the ultimate fate of these growing waves because it only captures phenomena at small wave amplitudes close to the uniform state. Therefore, the technique of linear stability analysis alone cannot explain the question of why the disturbance waves can develop into bubbles in gas-fluidized beds, but fail to form bubbles in liquid-fluidized beds.

In order to better understand the cause of bubbling, nonlinear terms in the equations of motion have been taken into account (Ding and Gidaspow, 1990; Fanucci et al., 1979; Hayakawa et al., 1994; Hirayama and Takaki, 1997; Liu, 1983; Needham and Merkin, 1983, 1986; Sasa and Hayakawa, 1992). Fanucci et al. (1979) analyzed the equations of motion based on the method of characteristics. They showed an occurrence of a shock discontinuity and speculated that the shock fronts physically manifest as bubbles. Liu (1983) used the principles of wave-hierarchy to explain certain wave interaction features in liquid fluidized beds. Needham and Merkin (1983; 1986) established a quasi-steady periodic state when the bed was unstable, based on a nonlinear dynamics analysis. They showed that this quasi-steady periodic state was stable only under a limited range of conditions. Although the investigation of the nonlinear equations yielded many interesting insights into the physics governing the behavior of the fluidized beds, they failed to reveal the mechanism of bubbling and non-bubbling behaviors in gas and liquid fluidized beds because they only focused on motion in one-dimension.

Direct integration of the volume-averaged equations of motion for gas-particle fluidized beds has been carried out in several papers (Anderson et al., 1995; Gidaspow et al., 1986; Kuipers et al., 1991). Anderson et al. (1995) added a small perturbation to the fully developed primary waves and carried out direct numerical integration of the equations of motion. They found that a rising bubble developed in a gas-fluidized bed with 200μm diameter glass beads in air, while in a liquid-fluidized bed with 1mm diameter glass beads in water the secondary instability developed very slowly with relatively small variations of density amplitude.

The approach of bifurcation analysis was also used to investigate the waveforms in fluidized beds (Dankworth and Sundaresan, 1991; Glasser et al., 1996, 1997; Glasser et al., 1998; Goz, 1992, 1993, 1995). Dankworth and Sundaresan (1991) examined the instability of a uniformly fluidized bed using nonlinear dynamics. They found that a new branch of plane or one-dimensional solutions could be born at a Hopf bifurcation point, at which the uniformly fluidized bed had a transition from a linearly stable state to a linearly unstable state. Göz (1992) used a bifurcation approach to investigate the instability of fluidized beds. He showed effects of various parameters on the evolution of traveling waves from uniformly fluidized beds. By investigating the development of lateral structure in one-dimensional vertical waves, Göz (1995) showed that a secondary instability could arise and the ratio of the growth rates of the primary and secondary instabilities could determine the difference between gas and liquid beds. Glasser et al. (1996; 1997) used bifurcation analysis and numerical continuation to investigate the stability features of fluidized beds. Their computational exploration showed how one–dimensional traveling wave solutions bifurcate from the uniform state and how this primary wave train is unstable to transverse waves with the variation of physical parameters and wave numbers. A connection between bubbles and clusters was established by Glasser et al. (1998), based on the same bifurcation approach. It was shown that bubbles and clusters belong to the same family of non-uniform solutions.

It is worth noting that only unbounded fluidized beds were considered in the above stability studies. In the present work, the technique of hydrodynamic stability analysis will be extended to bounded fluidized beds in order to better understand their stability behaviors.

1.5 Experimental Studies in Fluidized Beds

In order to test the applicability of the theoretical models and the constitutive relations, properly designed experiments are required. It is widely recognized that particle characteristics have a significant effect on the behavior of gas-fluidized beds. Therefore, a fundamental understanding of the behavior of granular materials in the fluidized state is key to the development of predictive models.

It has been shown that the different flow behaviors in fluidized beds are most likely the result of varying interparticle forces (Agbim et al., 1971; Geldart et al., 1984; Geldart and Wong, 1987; McLaughlin and Rhodes, 2001). It is widely accepted that the distinction between the bubbling and non-bubbling behavior in the fluidized beds is due to the contribution of a term involving the particle phase stress tensor (Grag and Pritchett, 1975). Normally, the particle phase stress tensor is composed of a particle pressure part and a particle viscosity part, which are functions of the granular temperature in kinetic theory.

The particle pressure is the pressure that is exerted by the particles on an imaginary plate inserted into the flow. Polashenski and Chen (1997) used a piezoelectric pressure transducer to measure the instantaneous and the long-time particle pressure, and calculated the granular temperature based on their measurement results and the constitutive relations from kinetic theory. They found that the solids viscosity decreases and the granular temperature increases as the superficial gas velocity is increased. Zenit et al. (1997) measured the particle pressure in a liquid fluidized bed and showed that the particle pressure critically depends on the solids volume fraction. A maximum in the particle pressure was found at an intermediate value of the solids fraction between zero and the packed bed limit.

One difficulty in testing the applicability of kinetic theory in the fluidized beds is that the particle phase closures depend on the granular temperature and this parameter is difficult to measure. As discussed earlier in this chapter, the granular temperature is related to the mean fluctuation velocity. Therefore, the value of this quantity can be obtained by measuring the displacement of particles. Menon and Durian (1997) used diffusing wave spectroscopy to measure the granular temperature in gas-particle fluidized beds. The regime of their measurement is limited to a few particle diameters inside the bed walls. They found the velocity fluctuation is very small in the uniformly fluidized state and the bed behaves as a weak solid. Cody et al. (1996) introduced and validated a novel non-intrusive probe to determine the granular temperature of the particles at the walls of a gas fluidized bed. By measuring the acoustic shot noise excitation of the surface of the walls by random particle impact, the particle normal velocity can be calculated at the walls, which can serve as critical boundary conditions for validating computational studies and numerical simulations. This technology was applied to a fluidized bed catalytic cracking unit by Cody et al. (2000), and they showed that the acoustic shot noise probe could detect changes in the chemistry and physics of gas generation and particle motion in the feed injection zone of the unit.

An experimental investigation on the effective drag force in a fluidized bed was presented by Makkawi and Wright (2003). The drag force on a particle was calculated based on the measurements of particle velocity and solids fraction during pulse gas tests, using twin-plane electrical tomography. They remarked on the validity of using $(1-\phi)^n$ for the reliable estimation of the effective drag force and showed that the parameter n is strongly dependent on the relative particle Reynolds number and the calculated drag force is comparable to the particle weight.

The flow profiles of fluidized beds have been presented in a large number of studies. In most gas-particle fluidized bed experiments, the solid flow was in the "Core-Annular" regime with a dilute rising core and a dense descending annular region, and the time-average solids flux exhibited a typically flattened parabolic shape with a maximum near the centerline (Benyahia et al., 2000; Fan et al., 2002; Helland et al., 2000; Hoomans et al., 2001; Miller and Gidaspow, 1992; Rhode and Laussmann, 1992; Samuelsberg and Hjertager, 1996; Wachem et al., 2001). The technology of Phase Doppler Particle Analyser was used to measure the solid mass flux and investigate the segregation phenomenon in circulating fluidized beds (Moortel et al., 1998; Moortel et al., 1997). Then a modified laser Doppler velocimetry method was developed to study concentrated suspension flows in a narrow rectangular channel (Koh et al., 1994; Lyon and Leal, 1998). By measuring fully developed particle velocity and concentration profiles, as well as the granular temperature in a narrow rectangular channel, Lyon and Leal (1998) found the particle concentration had a maximum near the centerline and a minimum at the channel walls, and the granular temperature had a sharp maximum at a position approximately 80% of the way from the channel axis towards the walls. Srivastava et al. (1998) used electrical capacitance tomography to image particle distribution, pressure profile and aeration rate in a circulating fluidized bed.

The first attempt to experimentally measure the properties of growing waves was done by Anderson and Jackson (1969). Subsequently, more systematic experimental studies were established by El-Kaissy and Homsy (1976). In their experiments, a one-dimensional wave train was observed and grew exponentially along the vertical direction for some distance above the distributor. This one-dimensional wave could become saturated, and a secondary instability could generate from the breakdown of the one-dimensional wave train if the amplitude of the one-dimensional waves was sufficiently large. The fact that the one-dimensional wave train is unstable to transverse structures under certain conditions indicates that the origin of bubbling may be related to this secondary instability.

It has been shown that the particle size distribution plays an important role in gas-particle mixtures. Many researchers have studied the effect of the particle size distribution on the hydrodynamic behavior of gas-fluidized beds. Grace and Sun (1991) studied the influence of fine particles in fluidized beds. They found that the minimum fluidization velocity was much higher in bimodal systems than mono-disperse particle systems. Wu and Baeyens (1998) experimentally examined the effect of a bimodal mixture in fluidized beds, and defined a mixing index whose value goes from 0 representing no mixing to 1 representing perfect mixing. They showed that the mixing index increases with an increase in gas superficial velocity. Rasuel et al. (1999) studied different types of segregation occurring in fluidized beds with a bimodal mixture. They found that at a low gas velocity the small, heavy particles are preferentially segregated at the bottom of the fluidized bed and large, light particles are at the top, while at a high gas velocity "inversion" can occur.

1.6 Outline

Chapter 2 introduces a continuum model to examine the gas-particle flows in a channel. The volume-averaged governing equations are used to describe the motion of the solid and the gas phases. The constitutive relations in the solid phase are obtained from kinetic theory. The boundary conditions are obtained by calculating the force balance and energy balance in a thin layer adjacent to the walls.

In Chapter 3, steady state solutions to equations of continuity and motion with closures from kinetic theory are sought and the variation of the solutions with changes in parameters are computed. The effect of physical parameters and operating conditions on the flow profiles are examined, using a bifurcation/continuation approach. The goal of this work is to understand and identify the origin of the sensitivity or insensitivity of the gas-particle flows to the physical or operating parameters (elasticity, gravity, bed width, solids fraction, the Reynolds number and the Froude number) from computational studies.

The instability of the steady state solutions is investigated in Chapter 4, using a linear stability analysis approach. The effects of elasticity, solids fraction and wall conditions on the instability structures are studied. The physical mechanism of the instability behaviors is investigated using a term-by-term analysis. Finally, a chain reaction is assumed in a gas-particle fluidized bed to study the effect of flow hydrodynamics on the chemical reactions.

In Chapter 5, we extend our investigation to granular and gas-particle flows with a binary particle mixture. Two models based on kinetic theory are examined to compare their solutions for steady fully developed flows. A parametric analysis of material properties and process parameters is carried out to study their influences on the flow profiles and the mixture behaviors. The physical mechanism of the segregation phenomenon is investigated based on a diffusion force analysis.

We conclude our work in Chapter 6, and discuss future avenues for investigation. For the progress of fluidization technologies, trial-and-error and empericism must be replaced by reliable and quantitative predictive methods to account for the flow instabilities and particle segregation.

Chapter 2 Equations of Motions

In this chapter, continuum models are established to describe gas-particle and granular flows, based on kinetic theory.

2.1 Equations of Motion for Gas-Particle Flows with Monodisperse Particles

Fluidized beds are typically modeled as two interpenetrating continua, representing the gas phase and solids phase respectively (Jackson, 1971, 1997, 2000). The geometry of the system considered is shown in Fig 2.1, where the gas flows from bottom to top and the particles are suspended due to the drag force. Here, 2Δ represent the width of the channel.

2.1.1 Governing Equations

The continuity and momentum balance equations for the gas and particle phases, originally proposed by Anderson & Jackson (1967), are used in the present study.

$$\frac{\partial \phi}{\partial t} + \nabla \cdot (\phi \underline{v}) = 0 \tag{2.1}$$

$$\frac{\partial (1-\phi)}{\partial t} + \nabla \cdot [(1-\phi)\underline{u}] = 0 \tag{2.2}$$

$$\rho_s \phi [\frac{\partial \underline{v}}{\partial t} + \underline{v} \cdot \nabla \underline{v}] = -\nabla \cdot \underline{\underline{\sigma}}_s - \phi \nabla \cdot \underline{\underline{\sigma}}_f + \underline{f} + \phi \rho_s \underline{g} \tag{2.3}$$

$$\rho_f (1-\phi) [\frac{\partial \underline{u}}{\partial t} + \underline{u} \cdot \nabla \underline{u}] = -(1-\phi) \nabla \cdot \underline{\underline{\sigma}}_f - \underline{f} + (1-\phi) \rho_f \underline{g} \tag{2.4}$$

where ϕ denotes the volume fraction of particles; $\underline{v}$ and $\underline{u}$ denote the local average velocities of particles and gas respectively; ρ_s and ρ_f are the solid and gas densities respectively; $\underline{g}$ is the specific gravity force; f is the average interaction force per unit volume of the bed, exerted on the particles by the gas; $\underline{\underline{\sigma}}_s$ and $\underline{\underline{\sigma}}_f$ are the effective stress tensors for the particle and gas phase respectively. Equation 2.1 and 2.2 represent the mass balance of the solid and gas phases, while Equation 2.3 and 2.4 represent the momentum balance of the solid and gas phases. In this model, the gas is assumed incompressible.

Since the solid phase stress depends on the granular temperature in kinetic theory, an equation representing the balance of pseudo-thermal energy (PTE) is required. This is obtained by subtracting the mechanical energy equation from the total energy equation.

$$\frac{\partial(\frac{3}{2}\rho_s\phi T)}{\partial t}+\nabla\cdot(\frac{3}{2}\rho_s\phi T\underline{v})=-\nabla\cdot\underline{q}-\underline{\underline{\sigma}}_s^c:\nabla\underline{v}-J_{coll}-J_{vis} \tag{2.5}$$

Equation 2.5, initially proposed by Agrawal et al. (2001), represents the PTE balance, where T denotes the granular temperature, which is proportional to the mean square of the particle fluctuation velocity ($T=\frac{1}{3}<\underline{C}\underline{C}>^2$, where $\underline{C}$ is the particle fluctuation velocity. $\underline{C}$=$\underline{c}$-$\underline{v}$, where $\underline{c}$ is the velocity of an individual particle.). The two terms on the left hand side represent the rate of accumulation and the rate of transport of PTE. On the right hand side, the first term represents the diffusive transport of PTE, the second term represents the rate of production of PTE by particle-particle shear, the third term denotes the rate of dissipation of PTE through inelastic collisions, and the final term denotes the viscous damping and the production of PTE by gas-particle slip.

2.1.2 Constitutive Relations

To solve this system, physically meaningful closures are required to close the above equations. In this model, the constitutive relations can be divided into two categories. The terms in the first category are related to the contribution of the gas phase, while the terms in the second category are only dependent on the interparticle interactions.

2.1.2.1 Constitutive Relations Related to the Gas Phase

The terms in this category stem from the existence of the gas phase, including the gas phase stress tensor $\underline{\underline{\sigma_f}}$, the gas-particle interaction force $\underline{f}$, and the contribution of the gas-particle slip on the fluctuational energy transport J_{vis}.

A form analogous to that of a Newtonian fluid is assumed for the gas phase stress tensor (Anderson and Jackson, 1968).

$$\underline{\underline{\sigma_f}} = P_f \underline{\underline{I}} - \mu_{eg}[\nabla \underline{u} + (\nabla \underline{u})^T - \frac{2}{3}(\nabla \cdot \underline{u})\underline{\underline{I}}] \tag{2.6}$$

where P_f is the gas phase pressure, and μ_{eg} is the effective gas viscosity. In this model, we adopt the expression for μ_{eg} suggested by Sinclair and Jackson (1989),

$$\mu_{eg} = \mu_g(1 + 2.5\phi + 7.6\phi^2)(1 - \phi / \phi_{\max}) \tag{2.7}$$

whose value vanishes when the particles are close packed with the maximum solids fraction ϕ_{max}, and equals the pure gas viscosity when the bed is operated in a very dilute condition.

Since only gas fluidized beds are considered in this work, virtual mass effects can be ignored and the interaction force is taken to have the following form (Anderson and Jackson, 1968).

$$\underline{f} = (1-\phi)\beta(\underline{u}-\underline{v}) \tag{2.8}$$

We have adopted the form for the drag coefficient, β, used by Anderson and Jackson (1968) for the bulk of our work,

$$\beta = \frac{\phi(\rho_s - \rho_f)g(1-\phi)^{1-N}}{v_t} \tag{2.9}$$

where v_t is the particle terminal velocity. This form is deduced from the well-known Richadson-Zaki (1954) correlation and represents a convenient form for a fluidized bed at low Reynolds numbers, and is consistent with the form adopted by Sinclair and Jackson (1989). In Equation (2.9), the exponent N depends on the Reynolds number based on the terminal velocity. Considering the normal flow velocity of fast fluidized beds, a value of 3 is used in the simulations for the exponent N (Richardson and Zaki, 1954). The evaluation of v_t was from the following correlation from Kunii and Levenspiel (1991).

$$d^* = d[\frac{\rho_f(\rho_s - \rho_f)g}{\mu_g}]^{1/3}$$

$$v_t^* = [\frac{18}{(d^*)^2} + \frac{0.591}{(d^*)^{0.5}}]^{-1}$$

$$v_t = v_t^*[\frac{\rho_f^2}{\mu_g(\rho_s - \rho_f)g}]^{-1/3} \tag{2.10}$$

where d is the particle diameter.

The fourth term on the right hand side of Equation 2.5, representing the contribution of the gas-particle slip on the fluctuational energy transport, was given by Agrawal et al. (2001).

$$J_{vis} = 3\beta T - \frac{81\phi\mu_g^2(\underline{u}-\underline{v})^2}{g_0 d^3 \rho_s \sqrt{\pi T}} \tag{2.11}$$

Unlike most previous work, in which the contribution of this term was not considered (Ocone et al., 1993; Pita and Sundaresan, 1991; Sinclair and Jackson, 1989), the effect of this term is investigated in the present study.

2.1.2.2 Constitutive Relations Dependent on the Interparticle Interactions

The mechanical behavior of a non-cohesive granular flow results from forces exerted at the contact points between different particles. These contacts are not permanent, but are continually forming and breaking as particles move. When the assembly of particles is widely distributed and in vigorous motion, individual contacts are of short duration and are treated as collisions. If the particles are highly packed (solids fraction > 0.5) (Johnson and Jackson, 1987; Johnson et al., 1990), on the other hand, the contacts are sustained contacts and the associated tangential frictional force becomes important. Since in our studies the average solids fraction is normally below 0.5, the contribution of the frictional stress is not considered and collisions between particles provide the principal mechanism for the transport of solids momentum and energy.

In the governing equations, the constitutive relations in the solids phase include the solids phase stress $\underline{\underline{\sigma_s}}$, the pseudo-thermal energy flux $\underline{q}$, and the pseudo-thermal energy dissipation J_{coll}. To describe these terms, a kinetic theory model, applied to rapid granular flows with smooth nearly elastic sphere particles, is used (Lun et al., 1984).

The closure for the particle stress takes the form

$$\underline{\underline{\sigma_s}} = \underline{\underline{P_k}} + \underline{\underline{P_c}} \tag{2.12}$$

$$\underline{\underline{P_k}} = \rho_s \phi T \underline{\underline{I}} - \frac{2\mu}{\eta(2-\eta)g_0}[1+\frac{8}{5}\eta(3\eta-2)\phi g_0]\underline{\underline{S}} \tag{2.13}$$

$$\underline{\underline{P_c}} = \rho_s 4\eta\phi^2 g_0 T \underline{\underline{I}} - \frac{16\mu\phi}{5(2-\eta)}[1+\frac{8}{5}\eta(3\eta-2)\phi g_0]\underline{\underline{S}} - \mu_b\eta[\frac{6}{5}\underline{\underline{S}} + (\nabla\cdot\underline{u})\underline{\underline{I}}] \tag{2.14}$$

where $\underline{\underline{P_k}}$ and $\underline{\underline{P_c}}$ represent the kinetic and collisional contributions to the total particle stress tensor; g_0 is the radial distribution function taken from Johnson and Jackson (1987), which constrains the solids fraction to remain less than the maximum packing limit.; μ is the viscosity of the particle phase; μ_b is the bulk viscosity for the particles and $\underline{\underline{S}}$ represents the deviatoric part of the particle phase rate of deformation tensor.

$$\mu = \frac{5\rho_s d\sqrt{\pi T}}{96} \qquad \mu_b = \frac{256\mu\phi^2 g_0}{5\pi} \qquad g_0 = \frac{1}{1-(\phi/\phi_{\max})^{1/3}}$$

$$\underline{\underline{S}} = \frac{1}{2}[\nabla\underline{v} + (\nabla\underline{v})^T] - \frac{1}{3}\nabla\cdot\underline{v}\underline{\underline{I}} \qquad \eta = \frac{1+e_p}{2} \tag{2.15}$$

where e_p is the coefficient of particle-particle restitution.

The pseudo-thermal energy flux is composed of two parts,

$$\underline{q_k} = -\frac{\lambda}{g_0}[(1+\frac{12}{5}\phi\eta^2 g_0(4\eta-3)\nabla T + \frac{12}{5}\eta(2\eta-1)(\eta-1)\frac{d}{d\phi}(\phi^2 g_0)\frac{T}{\phi}\nabla\phi] \tag{2.16}$$

$$\underline{q_c} = -\frac{12}{5}\lambda\phi\eta[(1+\frac{12}{5}\phi\eta^2 g_0(4\eta-3) + \frac{16}{15\pi}(41-33\eta)\phi\eta g_0]\nabla T$$

$$-\frac{12}{5}\lambda\phi\eta[\frac{12}{5}\eta(2\eta-1)(\eta-1)\frac{d}{d\phi}(\phi^2 g_0)\frac{T}{\phi}\nabla\phi] \tag{2.17}$$

$$\underline{q} = \underline{q_k} + \underline{q_c} = -\frac{\lambda}{g_0}[(1+\frac{12}{5}\phi\eta g_0)(1+\frac{12}{5}\phi\eta^2 g_0(4\eta-3)+\frac{64}{25\pi}(41-33\eta)\phi^2\eta^2 g_0^2]\nabla T$$

$$-\frac{\lambda}{g_0}[(1+\frac{12}{5}\phi\eta g_0)\frac{12}{5}\eta(2\eta-1)(\eta-1)\frac{d}{d\phi}(\phi^2 g_0)\frac{T}{\phi}\nabla\phi \qquad (2.18)$$

where $\underline{q_k}$ and $\underline{q_c}$ represent the kinetic and collisional contributions to the total pseudo-thermal energy flux; λ is the pseudo-thermal conductivity:

$$\lambda = \frac{75\rho_s d\sqrt{\pi T}}{48\eta(41-33\eta)} \qquad (2.19)$$

Our simulation shows that the second group of terms on the right hand side of the expression of $\underline{q}$, which denotes the contribution to the flux of the pseudo-thermal energy is proportional to the gradient of the particle concentration, is very small relative to the other terms, even when the coefficient of particle-particle restitution is smaller than 0.9. Thus this term is neglected in the calculations.

The term representing the pseudo-thermal energy dissipation takes the form

$$J_{coll} = \frac{48}{\sqrt{\pi}}\eta(1-\eta)\frac{\rho_s \varepsilon_s^2}{d} g_0 T^{\frac{3}{2}} \qquad (2.20)$$

It is clear that this term vanishes when the particles are assumed elastic.

2.1.3 Boundary Conditions

The boundary conditions for granular flows have been investigated in a number of papers (Hui et al., 1984; Jenkins and Richman, 1986; Johnson and Jackson, 1987; Richman, 1988). In this work, the boundary conditions proposed by Johnson and Jackson (1987) are used (see Fig 2.2). In their model the boundary conditions are dependent on two wall properties, the coefficient of restitution for particle-wall collisions e_w, and the specularity coefficient φ' which represents the roughness of the wall. The value of φ' ranges from zero, if the collisions between the particles and walls are specular, to unity if the incident particles are scattered diffusely. Although Jenkins and Richman (1986) characterized the wall properties in a more detailed manner, it is found that if the specularity coefficient in the Johnson and Jackson (1987) model can be represented in terms of the details of wall structures, these two models essentially have the same form (Alam and Nott, 1998).

By equating the tangential stress exerted by the flowing particles in the bulk adjacent to the solid boundary with the tangential momentum flux due to particle-wall collisions, the condition on the slip velocity has the form

$$\frac{\underline{u_{sl}} \cdot \underline{\underline{\sigma_s}} \cdot \underline{n}}{|\underline{u_{sl}}|} + \frac{\varphi' \sqrt{3} \pi \rho_s \phi T^{1/2} |\underline{u_{sl}}|}{6 \phi_{max} (1 - (\phi / \phi_{max})^{1/3})} = 0 \tag{2.21}$$

where $\underline{u_{sl}}$ is the slip velocity between the wall and the particles in contact with it, and $\underline{n}$ is the unit normal vector directed from the boundary into the particle assembly. In Equation 2.21, the second term, representing the rate of momentum transfer to a unit area of the wall surface by the particle-wall collisions, is the product of the average collision frequency for each particle, $(3T)^{\frac{1}{2}} / [d(\phi_{max} / \phi)^{\frac{1}{3}} - 1]$, the average tangential momentum transferred per collision, $\varphi' \pi \rho_s \phi d^3 u_{sl} / 6$, and the number of particles adjacent to a unit area of the wall surface, $[d^2 (\phi_{max} / \phi)^{\frac{2}{3}}]^{-1}$.

The second boundary condition is a statement of the pseudo-thermal energy balance in a slice of small thickness enclosing the solid boundary, where the energy flux normal to the wall is balanced by the net production of energy at the wall:

$$\underline{n}\cdot\underline{q}+\gamma^{w}-\underline{v_{sl}}\cdot\left[\frac{\varphi'\sqrt{3}\pi\phi T^{1/2}}{6\phi_{\max}[1-(\phi/\phi_{\max})^{1/3}]}\right]\underline{v_{sl}}=0 \tag{2.22}$$

In Equation 2.22, the first term represents the flux of pseudo-thermal energy to the walls, the second term is the energy dissipation rate per unit area due to inelastic particle-wall collisions, and the third term represents the generation of pseudo-thermal energy due to particle-wall slip. The dissipation term is the product of the average energy loss per particle-wall collision, $\frac{1}{4}\pi\rho_s d^3 T(1-e_w^2)$, the average collision frequency per particle, and the number of particles adjacent to a unit area of the boundaries.

$$\gamma^{w}=[\frac{1}{4}\pi\rho_s T(1-e_w^2)]\frac{\phi(3T)^{1/2}}{\phi_{\max}[1-(\phi/\phi_{\max})^{1/3}]} \tag{2.23}$$

In most theoretical studies of gas-particle fluidized beds, a no-slip or (stress) free boundary condition is applied to the gas phase. This is true for the point velocity of the gas at the walls. However, the gas velocity appearing in the equations of motion is not this, but some suitable average. Therefore, an appropriate boundary condition for the gas velocity at the walls is a subject of debate. The boundary condition representing the force balance of the gas phase originally proposed by Sinclair and Jackson (1989) was used in this work,

$$(\beta(\underline{u}-\underline{v})+\frac{\partial \underline{\underline{P}}}{\partial \underline{x}})\delta+\mu_{eg}\underline{n}\cdot\nabla\underline{u}+\frac{2\mu_{eg}T\underline{u}}{v_t^2\delta}=0 \tag{2.24}$$

where δ is a parameter related with the thickness of a layer adjacent to the wall in which the gas phase force balance is calculated. In Equation 2.24, the first term represents the drag force and the force due to the axial gradient of the gas pressure, the second term is the shear stress on the surface of the layer remote from the wall, the last term is a simple description of the shear stress at the wall. Using this boundary condition, the no-slip boundary condition and the Poiseuille solution for gas flows can be observed when the solids fraction is close to zero. When the solids load is close to the packed limit, the last two terms vanish, leading to a force balance in a bed of infinite extent.

2.1.4 Comparison with Previous Models

In previous work, simpler models have been used to simulate the hydrodynamics in gas-fluidized beds (Anderson et al., 1995; Anderson and Jackson, 1967, 1968; Glasser et al., 1996, 1997), where the solids phase stress was not dependent on the granular temperature. Therefore, the PTE balance equation was not required. To investigate the contribution of the PTE equation on bounded gas-particle flows, the model, originally proposed by Glasser, et al. (1997), is examined in a channel with no-slip boundary conditions for both gas and particle phases. In this model, the dimensional equations for a fully developed flow take the following form:

$$\frac{dp_f}{dx} + \mu_{eg}\frac{d^2u_x}{dy^2} - \beta(u_x - v_x) - \rho_g g = 0 \tag{2.25}$$

$$\frac{d}{dy}(\mu_s \frac{dv_x}{dy}) + \beta(u_x - v_x) - (\rho_s - \rho_g)g\phi = 0 \tag{2.26}$$

$$\frac{d}{dy}(p_s) = 0 \tag{2.27}$$

where μ_{eg} is given by Equation 2.7, β is given by Equation 2.9, $p_s = C_0\phi^3 \exp[(r\phi)/(\phi_{\max} - \phi)]$, and $\mu_s = A_1\phi/[1-(\phi/\phi_{\max})^{1/3}]$. Here, C_0, r, A_1 are constants.

We have gone on to compute the solids fraction, and gas and particle velocity profiles for these equations. Although the gas and solid velocity profiles, computed using the model adopted by Glasser et al. (1997) model share similar features to those computed using the present model, the solids fraction is always uniform across the bed because Equation 2.27 can be simplified as ϕ =constant. Clearly, the uniform solids fraction profile is unrealistic for bounded channel flows, so the PTE equation is indispensable.

2.2 Equations of Motion for Granular Flows with Bidisperse Particles

It is well known that multiple components and particle size distributions clearly complicate flow behaviors. Segregation occurs if particles differ in size, density, shape, or other characteristics. A natural starting point to understand the effect of the poly-dispersity on the flow behavior and mixture rheology is an investigation of a bidisperse system of smooth and slightly inelastic spheres.

In this section, we extend our work to involve a binary particle mixture. To make the analysis simpler, the effect of the gas phase is not considered. The governing equations as well as boundary conditions for granular flows with a binary particle mixture are investigated, using the kinetic theory. The contribution of the gas will be considered in the next section.

2.2.1 Governing Equations

In the present work, species A represents large/heavy particles with diameter d_A and mass m_A, while species B represents small/light particles with diameter d_B and mass m_B.

To study the significance of the non-equipartition of the granular energy, two different models are developed. The first model (JM model) is based on the work of Jenkins and Mancini (1989) with a modification by Aranson and Willits (1998), where the single velocity distribution function has a non-Maxwellian form and the assumption of equipartition of granular energy is used. The second one (IA model) is based on the work of Iddir and Arastoopour (2005), in which non-equipartition of granular energy is taken into account and the velocity distribution has a non-Maxwellian form.

In the JM model, the continuity equation of each particle phase is

$$\frac{\partial \phi_i}{\partial t} + \nabla \cdot (\phi_i \underline{v_i}) = 0 \tag{2.28}$$

where ϕ_i and $\underline{v_i}$ denote the solids fraction and the average velocity of particle species i. The momentum balance for each particle phase is

$$\rho_i \phi_i [\frac{\partial v_i}{\partial t} + \underline{v_i} \cdot \nabla \underline{v_i}] = -\nabla \cdot \underline{\underline{P_{i-JM}}} + n_i \underline{F_i} + \underline{\psi_i} \tag{2.29}$$

where ρ_i is the density of particle species i; $\underline{F_i}$ denotes external forces on the particle species i; $\underline{\underline{P_{i-JM}}}$ is the stress tensor and $\underline{\psi_i}$ represents the momentum source results from collisions between unlike particles. Since the assumption of equipartition of granular energy is used in this model, the balance of the mixture pseudo-thermal energy becomes

$$
\begin{aligned}
&n(\frac{\partial T}{\partial t}+\underline{v}\cdot\nabla T)-T\nabla\cdot(n_A\underline{v_A'}+n_B\underline{v_B'})-\sum_{i=A,B}F_i\cdot(n_A\underline{v_A'}+n_B\underline{v_B'})= \\
&-\nabla\cdot\underline{q_{JM}}-\underline{\underline{P_{JM}}}:\nabla\underline{v}-\gamma
\end{aligned}
\tag{2.30}
$$

where $\underline{q_{JM}}$ is the mixture energy flux; $\underline{v}$ is the mixture average solids velocity, $\underline{v}=\frac{n_A m_A \underline{v_A}+n_B m_B \underline{v_B}}{n_A m_A+n_B m_B}$; T is the mixture granular energy, $T=\frac{n_A T_A+n_B T_B}{n_A+n_B}$, where T_i represents the granular energy of particle species i, $T_i=\frac{1}{3}m_i<C_iC_i>$ where C_i is the fluctuation velocity of particle species i; n_i represents the number density and n is the total number density, $n=n_A+n_B$; $\underline{\underline{P_{JM}}}$ is the mixture solids stress tensor, $\underline{\underline{P_{JM}}}=\underline{\underline{P_{A-JM}}}+\underline{\underline{P_{B-JM}}}$; γ is the rate of granular energy dissipation; $\underline{v_i'}$ denote the diffusion velocity of particle species i. The expressions of $\underline{\underline{P_{i-JM}}}$, $\underline{\psi_i}$, $\underline{q_{JM}}$ and γ can be found in Appendix 1.

In the IA model, the continuity equation is same as the JM model (see Equation 2.28). The momentum balance for each particle phase takes the form

$$
\rho_i\phi_i[\frac{\partial v_i}{\partial t}+\underline{v_i}\cdot\nabla\underline{v_i}]=-\nabla\cdot\underline{\underline{P_i}}+\frac{\phi_i\rho_i}{m_i}\underline{F_{iext}}+\underline{F_{D1i}}+\underline{F_{D2i}}
\tag{2.31}
$$

where $\underline{F_{iext}}$ denotes external forces on the particle species i; and $\underline{F_{D1i}}$ and $\underline{F_{D2i}}$ represent the collisional momentum source. The pseudo-thermal energy balance for each particle phase is

$$
n_i(\frac{\partial T_i}{\partial t}+\underline{v_i}\cdot\nabla T_i)=-\nabla\cdot(\underline{q_{1i}}+\underline{q_{2i}})-\underline{\underline{P_i}}:\nabla\underline{v_i}+N_i-\underline{v_i'}\cdot(\underline{F_{D1i}}+\underline{F_{D2i}})
\tag{2.32}
$$

where $\underline{q_{1i}}$ and $\underline{q_{2i}}$ are the energy flux and N_i denotes the collisional energy dissipation. The constitutive relations are shown in Appendix 1.

2.2.2 Boundary Conditions

The boundary conditions at the walls for the particle phase momentum balance and pseudo thermal energy balance follow the equations of Johnson and Jackson (1987), and extend to the bidisperse model by considering the definition of the granular energy.

For the JM model, the boundary conditions for the particle phase momentum balance and pseudo thermal energy balance take the form

$$\frac{\underline{v_{sli}} \cdot \underline{\underline{P_{i-JM}}} \cdot \underline{k_i}}{|\underline{v_{sli}}|} + \frac{\varphi'\sqrt{3}\pi(\rho\phi T n)^{1/2}|\underline{v_{sli}}|}{6\phi_{\max}(1-(\phi/\phi_{\max})^{1/3})} = 0 \tag{2.33}$$

$$\underline{k} \cdot \underline{q_{JM}} = \frac{\pi(3Tn\rho\phi)^{1/2}}{6\phi_{\max}(1-(\phi/\phi_{\max})^{1/3})}[\varphi'|\underline{v_{sl}}|^2 - \frac{3Tn}{2\rho\phi}(1-e_w^2)] \tag{2.34}$$

where $\underline{k_i}$ is the unit normal vector directed from the boundary into the particle species i assembly; $\rho=(\phi_A\rho_A+\phi_B\rho_B)/(\phi_A+\phi_B)$; $\underline{v_{sl}}$ is the slip velocity between the mixture particle phase and the walls; $\underline{v_{sli}}$ is the slip velocity between the particle phase i and the walls. The definition of the parameters φ', $\phi_{\max}$ and e_w can be found in Section 2.2.

The boundary conditions of IA model are similar with the JM model except that the mixture granular energy should be replaced with the individual granular energy.

$$\frac{\underline{v_{sli}} \cdot \underline{\underline{P_i}} \cdot \underline{k_i}}{|\underline{v_{sli}}|} + \frac{\varphi'\sqrt{3}\pi(\rho_i\phi_i T_i n_i)^{1/2}|\underline{v_{sli}}|}{6\phi_{\max}(1-(\phi_i/\phi_{\max})^{1/3})} = 0 \tag{2.35}$$

$$\underline{k_i} \cdot \underline{q_i} = \frac{\pi(3T_i n_i\rho_i\phi_i)^{1/2}}{6\phi_{\max}(1-(\phi_i/\phi_{\max})^{1/3})}[\varphi'|\underline{v_{sli}}|^2 - \frac{3T_i n_i}{2\rho_i\phi_i}(1-e_w^2)] \tag{2.36}$$

2.3 Equations of Motion for Gas-Particle Flows with Bidisperse Particles

When gas is added into the system, particles may exhibit a fluidized state due to the drag force acted from the gas to the particle assembly. The continuity equations of the gas and particle phase are

$$\frac{\partial(1-\phi_t)}{\partial t}+\nabla\cdot[(1-\phi_t)\underline{u}]=0 \tag{2.37}$$

$$\frac{\partial\phi_i}{\partial t}+\nabla\cdot(\phi_i\underline{v_i})=0 \tag{2.38}$$

where ϕ_i is the solids fraction of particle species i and ϕ_t is the total solids fraction; $\underline{u}$ represents the gas phase velocity and $\underline{v}_i$ denotes the velocity of particle species i. The momentum balances for gas and particle phase are

$$\rho_f(1-\phi)[\frac{\partial u}{\partial t}+\underline{u}\cdot\nabla\underline{u}]=-(1-\phi)\nabla\cdot\underline{\underline{\sigma_f}}-\underline{f}+(1-\phi)\rho_f\underline{g} \tag{2.39}$$

$$\rho_i\phi_i[\frac{\partial v_i}{\partial t}+\underline{v_i}\cdot\nabla\underline{v_i}]=-\nabla\cdot\underline{\underline{P_i}}-\phi_i\nabla\cdot\underline{\underline{\sigma_f}}+\underline{f_i}+\frac{\phi_i\rho_i}{m_i}\underline{F_{iext}}+\underline{F_{D1i}}+\underline{F_{D2i}} \tag{2.40}$$

where $\underline{f}$ is the average interaction force per unit volume of the bed exerted on the total particles by the gas, and $\underline{f_i}$ represents this force on particle species i. The balance of pseudo-thermal energy is

$$n_i(\frac{\partial T_i}{\partial t}+\underline{v_i}\cdot\nabla T_i)=-\nabla\cdot(\underline{q_{1i}}+\underline{q_{2i}})-\underline{\underline{P_i}}:\nabla\underline{v_i}+N_i-\underline{v_i'}\cdot(\underline{F_{D1i}}+\underline{F_{D2i}}) \tag{2.41}$$

To close the above equations, the kinetic theory model proposed by Iddir and Arastoopour (2005) is used. The definition of $\underline{\underline{P_i}}$, $\underline{F_{iext}}$, $\underline{F_{D1i}}$, $\underline{F_{D2i}}$, $\underline{q_{1i}}$, $\underline{q_{2i}}$ and N_i can be found in Section 2.2.1, and their correlations are shown in Appendix 1. The gas-particle interaction force is taken to have the following form (Anderson and Jackson, 1968).

$$\underline{f_i} = (1-\phi_t)\beta_i(\underline{u} - \underline{v_i}) \tag{2.42}$$

For the drag coefficient, β_i, an ideal state is assumed so that the form, originally proposed by Anderson and Jackson (1968), can be adopted.

$$\beta_i = \frac{\phi_i(\rho_i - \rho_f)g(1-\phi_t)^{1-N}}{v_{ti}} \tag{2.43}$$

The evaluation of the terminal velocity of particle species i, v_{ti} was from the correlation from Kunii and Levenspiel (1991) (see Equation 2.10).

The boundary conditions at the wall for the particle phase momentum balance and pseudo thermal energy balance are taken from Equation 2.35 and 2.36. An appropriate boundary condition for the gas velocity, however, is a subject of debate. For the current model, a boundary condition representing the force balance of the gas phase in a thin layer adjacent to the wall should be derived. This is not trivial and some physical terms are still not completely understood for the binary particle system. From our work on gas particle flows with a mono-disperse particle mixture, however, we found that the effect of the gas boundary conditions is not significant. Simulations based on no-slip, (stress) free, and integration gas phase boundary conditions showed that the flow profiles are not sensitive to the gas phase boundary conditions (see Chapter 3). Therefore, a no-slip boundary condition is used for the gas phase.

$$\underline{u} = \underline{0} \tag{2.44}$$

Fig 2.1: Schematic diagram of a bounded gas-particle flow in a fluidized bed.

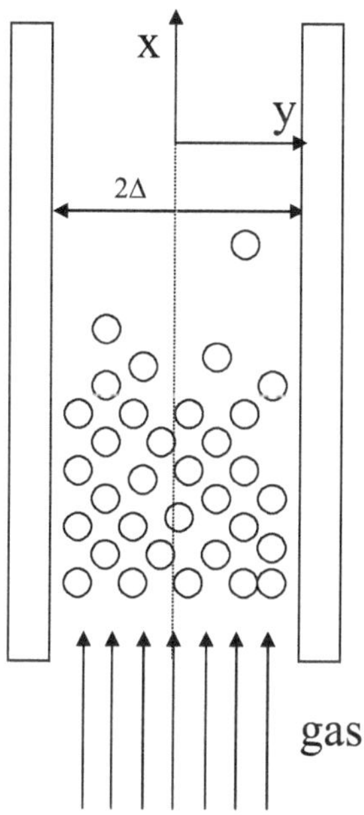

Fig 2.2: Schematic diagram for boundary conditions.

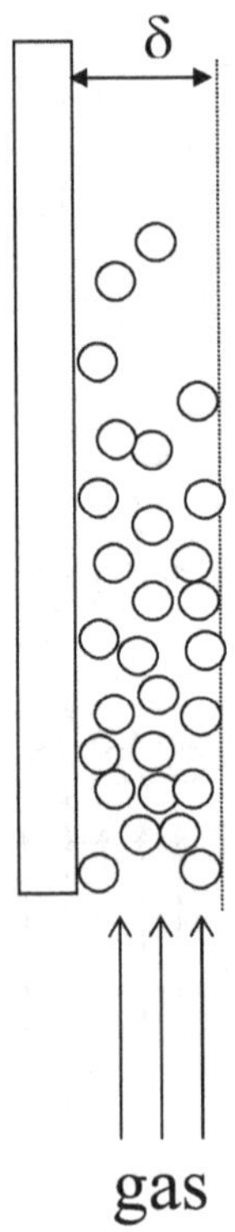

Chapter 3 Steady State Solutions for Bounded Gas-Particle Fluidized Beds with Monodisperse Particles

3.1 Model Equations and Numerical Method

For steady, fully developed flows, the continuity equation is identically satisfied. Using ρ_s, v_t, Δ and Δ / v_t as characteristic density, velocity, length and time, the momentum and pseudo-thermal energy equations (Equation 2.3 to 2.5) can be cast in dimensionless forms.

$$-\frac{BD}{C}\frac{dp_f^*}{dx^*} - \frac{BD(1-E)}{C} f_0(u_x^* - v_x^*) - \frac{BDE}{C} + \frac{d}{dy}(\mu_{eg}^* \frac{du_x^*}{dy^*}) = 0 \tag{3.1}$$

$$\frac{D(1-E)}{C}[f_0(u_x^* - v_x^*) - \phi] + \frac{d}{dy}(\frac{1}{2} f_3 (T^*)^{\frac{1}{2}} \frac{dv_x^*}{dy^*}) = 0 \tag{3.2}$$

$$\frac{d}{dy}(f_1 T^*) = 0 \tag{3.3}$$

$$\begin{aligned} &\frac{d}{dy}(f_4 (T^*)^{\frac{1}{2}} \frac{dT^*}{dy}) - D^2 f_5 (T^*)^{\frac{3}{2}} - \frac{(1-E)D}{C} 3 f_0 T^* + \frac{D^2}{B^2} f_2 (u_x^* - v_x^*)^2 (T^*)^{-\frac{1}{2}} \\ &+ f_3 (T^*)^{\frac{1}{2}} \frac{1}{2} (\frac{dv_x^*}{dy^*})^2 = 0 \end{aligned} \tag{3.4}$$

where the dimensionless parameters and dimensionless variables are

$$B = \frac{\rho_s d v_t}{\mu_g} \qquad C = \frac{v_t^2}{g\Delta} \qquad D = \frac{\Delta}{d} \qquad E = \frac{\rho_g}{\rho_s}$$

$$p_f^* = \frac{p_f}{\rho_s g \Delta} \qquad u_x^* = \frac{u_x}{v_t} \qquad v_x^* = \frac{v_x}{v_t} \qquad T^* = \frac{T}{v_t^2} \qquad \mu_{eg}^* = \frac{\mu_{eg}}{\mu_g}$$

The functions f_0 to f_5, which depend only on the solids volume fraction, are dimensionless and are given in Table 3.1. In the present study, the majority of the results are presented in dimensionless form.

The boundary conditions (Equation 2.21, 2.22, 2.24) take the following forms in terms of dimensionless variables

$$\frac{dv_x^*}{dy^*} + D\varphi' f_6 v_x^* = 0 \tag{3.5}$$

$$\frac{dT^*}{dy^*} - D\varphi' f_7 (v_x^*)^2 + D(1-e_w^2) f_8 T^* = 0 \tag{3.6}$$

$$\mu_{eg}^* \frac{du_x^*}{dy^*} + \frac{BDG(1-E)}{C}\frac{\phi}{\phi_{\max}} f_0 (u_x^* - v_x^*) + \frac{BDG}{C}\frac{\phi}{\phi_{\max}}\frac{dp_f^*}{dy^*} + 2G\frac{\phi_{\max}}{\phi} T^* u_x^* = 0 \tag{3.7}$$

where $f_6 - f_8$ are dimensionless functions defined in Table 3.1; G is a dimensionless variable related to the thickness of a layer adjacent to the wall in which we calculate the force balance. To study the effect of the gas boundary conditions on flow profiles, simulations based on no-slip and stress free gas phase boundary conditions have been also carried out. We found under our physical situation the flow profiles are not sensitive to the gas phase boundary conditions. Only a small difference can be observed in the region close to the walls (see Fig 3.1). However, to keep consistent with equations with averaged forms and previous work, the expression proposed by Sinclair and Jackson (1989) was used in the following simulations. It is found that the governing equations along with the boundary conditions admit the following symmetry

$$u_x^*(y) = u_x^*(-y) \qquad v_x^*(y) = v_x^*(-y) \qquad \phi^*(y) = \phi^*(-y) \qquad T^*(y) = T^*(-y)$$

Thus, a relevant boundary condition at y=0 can be written as

$$\frac{du_x^*}{dy^*}=0 \qquad \frac{dv_x^*}{dy^*}=0 \qquad \frac{dT^*}{dy^*}=0 \tag{3.8}$$

The base physical parameters used in this model are listed in Table 3.2, representing cocurrent upflow of 400μm diameter particles in air at 500°C and 6.5 atmosphere in a narrow channel. Equation 3.1 – 3.8 describes a steady state gas-particle laminar flow in a channel.

The above model consists of four ordinary differential equations and four unknowns (u, v, ϕ and T), subject to two point boundary conditions. Equation 3.3 can be rewritten as an algebraic equation to provide an explicit relation between ϕ and T. This eliminates one unknown and leaves only three differential equations and three unknowns. A numerical method based on the orthogonal collocation approach (Constantinides, 1987) is set up to obtain the solutions for the base state (see Appendix 2). Another numerical scheme, namely, a second-order finite difference method is also used to ascertain the correctness of the solutions. Preliminary results confirm that both numerical methods yield the same solutions (see Appendix 2).

From the simulation, it is found that multiple solutions can be obtained with different initial guesses. To find all possible solutions and track the variation of the solutions with different parameters, we make use of a bifurcation analysis and continuation approach with AUTO 86 software (Doedel et al., 1991). The calculation procedure of the continuation approach involves the following steps. First, input a steady state solution into the AUTO 86 software as the initial state. Then the software calculates the solution branches as a function of the chosen parameter as other physical parameters are kept unchanged. In the present work, the continuation is carried out based on two different algorithms. The first is examining the variation of the solids fraction amplitude, ϕ_{max}-ϕ_{min}, with a chosen parameter, which requires calculating the differential form of Equation 3.3 with two constraints, $d\phi / dy^* = 0$ at the center and the average solids fraction across the channel is a constant. The second involves examining the change of the average solids fraction with a chosen parameter. This requires rewriting Equation 3.3 as an algebraic form,

$$f_1 T^* = C_1 \tag{3.9}$$

and calculating the steady state solutions in terms of the coefficient C_1.

3.2 Results and Discussions

An overall view of possible flow regimes of gas-particle flow in a vertical channel is shown in Fig 3.2 and 3.3. This is examined by plotting the variation of the dimensionless gas volume flow rate Q^*_g as a function of the dimensionless solid volume flowrate Q^*_s for a given value of the pressure drop, where $Q_s^* = Q_s / (\Delta^2 v_t)$ and $Q_g^* = Q_g / (\Delta^2 v_t)$. In Fig 3.2, the curve in the first quadrant represents cocurrent upward flow, and the curves in the second and third quadrant describe countercurrent flow and cocurrent downward flow respectively. When the gas flowrate is reduced in the first quadrant, the solid flow rate increases initially, passes through a maximum and then decreases. After crossing the axis into the second quadrant, the solids flowrate continues decreasing, then passes through a minimum in the third quadrant and moves back towards positive values of the solid flow rate. Fig 3.2 clearly shows the existence of multiple solutions, which means that for a given solid flow rate, the system can have more than one steady state solution. This kind of contour map and steady-state multiplicity was also reported by Sinclair and Jackson (1989). It is interesting to note that for a zero value of the solid flowrate, the gas flow rate has two values, where the upper one represents a dilute fluidized state while the lower one represents a fluidized state close to the maximum particle packing. When the gas pressure drop is increased (see Fig 3.3), only the cocurrent upflow is exhibited in the flow regime. With a decrease in the gas flowrate, the solid flow rate increases initially, passes through a maximum and then decreases towards a fluidized state close to the maximum particle packing.

In this work, we have focused on solutions in the first quadrant, which corresponds to cocurrent upflow.

3.2.1 Effect of Inelasticity

The contribution of particle inelasticity on granular flows and gas-particle flows has been discussed in a number of papers (Goldhirsch and Zanetti, 1993; Hrenya and Sinclair, 1997; Tan and Goldhirsch, 1997). Fig 3.4 shows the solids fraction amplitude, ϕ_{max}-ϕ_{min}, as a function of the coefficient of particle-particle restitution. When e_p is close to 1, the solids fraction amplitude is very sensitive to this parameter. However, when its value is smaller than 0.985, a plateau can be observed, indicating that e_p has little effect on the flow structures. To better understand the structures of the branches in the continuation map, the flow profiles for points A, B, C, D, E, F are plotted. Fig 3.5 shows the variation of the solids fraction, granular temperature, and gas and solid velocities within the fluidized bed for Points A, B, E and F. When elastic collisions between particles in fluidized beds are assumed (Point A, e_p=1), there is a plateau of low solid density in the center region and a relatively large concentration of particles near the walls. The granular temperature increases slightly when moving away from the center, passes through a maximum and then decreases rapidly near the walls. The relation of the solids fraction and the granular temperature is predicted in terms of Equation 3.3 that the solid concentration should be small in regions where the granular temperature is large, and vice versa.

The variation of gas and solid velocity for the elastic case is shown in Fig 3.5(c) and (d) respectively. Both velocities are higher at the center and lower at the walls. The velocity difference between the gas phase and solid phase is associated with the well-known drag force. It has been shown that the drag force plays a key role in the formation of heterogeneous flow structures in gas-particle fluidized beds when collisions between particles are ideal (Li and Kuipers, 2003). This "Core-Annular" flow structure in gas-particle fluidized beds has been discussed in many papers and is typically observed in fast gas-fluidized beds (Bader et al., 1988; Benyahia et al., 2000; Mathiesen et al., 2000; Miller and Gidaspow, 1992). However, when collisions between particles are slightly inelastic (Point B, e_p=0.995), the flow profiles exhibit a different structure. Under this condition, there is a segregation of particles near the center and a dilute region near the walls (Fig 3.5(a)). This type of flow structure has been observed in computer simulations of gas-fluidized beds (Helland et al., 2000; Sinclair 1989). To the best of our knowledge, this flow structure has not been experimentally observed in gas-fluidized beds although it has been observed in liquid-fluidized beds (Koh et al., 1994; Lyon and Leal, 1998). The gas phase velocity initially remains a constant value when moving away from the center, passes through a maximum corresponding to the maximum in the granular temperature profile, and then decreases quickly near the walls (Fig 3.5(c)). This relation between the granular temperature and the superficial gas velocity has been observed in experiments (Cody et al., 1996; Poilashenski and Chen, 1997). When e_p=0.995, the solid velocity has a maximum value in the center, and decreases when moving towards the walls (Fig 3.5(d)). When the system becomes more inelastic (Point E, e_p=0.96, and Point F, e_p=0.95), the segregation of particles is more pronounced and a plug flow with maximum solids fraction can be seen near the center. In Fig 3.5(c) and 3.5(d), the relative velocity between gas and solids is quite large in the region close to the walls if collisions between particles are inelastic. This is due to the segregation of particles. For inelastic cases, most of particles segregate in the center and there is a dilute region close to the walls. Since the drag coefficient decreases with a decrease in solids fraction, the value of the drag coefficient is much larger in the center than close to the walls. Thus, the gas-solids relative velocity increases from the center to the walls. Similar flow profiles were also obtained by Sinclair (1989).

Scrutiny of Fig 3.4 and Fig 3.5 reveals that the sensitivity to the coefficient of restitution occurs only when e_p is close to 1. When the gas-particle system transits from an elastic case to a slightly inelastic one, the particles start to migrate from the walls to the center. When the solids concentration is equal at the center and the walls, the solids fraction amplitude goes through a minimum (see Fig 3.4, e_p=0.998). With continuously decreasing e_p, the solids fraction becomes higher at the center than at the walls, leading to an increase in the solids fraction amplitude. Once e_p is smaller than 0.985, a solids fraction close to the packed bed limit is observed at the center, indicating that a maximum solids fraction amplitude is approached. With a further decrease in e_p, the gas-particle flow profiles are not sensitive to this parameter. Another interesting phenomenon observed in the continuation is that the profile of solids fraction amplitudes has a "S" form bifurcation diagram when e_p is between 0.990 and 0.986, indicating the existence of multiple solutions for a fixed set of physical parameters. In order to better understand this bifurcation diagram, the solids fraction profiles are plotted for Points C and D (e_p=0.988). Fig 3.6 shows that the solution branch corresponding to D-E represents a state of maximum solids fraction at the center, while the solution branch corresponding to B-C represents a transition process of particle segregation from the walls to the center.

By examining the continuation map and the solids fraction profiles (see Fig 3.4 and Fig 3.5), we can see that a transition from "Core-Annular" structure to center segregation of particles occurs when e_p decreases from 1 to 0.99, and with a further decrease in e_p the flow profiles are not sensitive to this parameter. This is in agreement with the results of Goldhirsch (1999) that the granular temperature is related to $1-e_p^2$. It is clear that when e_p is changed from 0.99 to 0.999, the value of $1-e_p^2$ is changed 10-fold. However, when e_p is changed from 0.95 to 0.96, the value of $1-e_p^2$ varies by only 1.24-fold. Accordingly, it is not surprising that the extreme sensitivity exists in the gas-particle flow when e_p is close to 1, and this sensitivity is reduced when the system becomes more inelastic. From the above discussion, it seems that this sensitivity to e_p is not totally unrealistic, however, the physical mechanism leading to this phenomenon remains unresolved.

From Fig 3.5, it can be observed that the granular temperature near the walls increases as e_p decreases, indicating that the flow structures have a potential to switch from a situation where the wall behaves as a sink of PTE to a source of PTE. This variation can be related to the value of the coefficient of particle-wall restitution (e_w). The effect of e_w on the solids fraction and granular temperature is shown in Fig 3.7. It can be seen that e_w plays a small role in the solids fraction profiles. Small variations are obtained only in the region near the walls when e_w is increased from 0.6 to 0.99. The impact of e_w on the granular temperature is more pronounced. When e_w is sufficiently large, the wall is a source of fluctuational energy. The value of the granular temperature increases on moving towards the walls from the center. If e_w is small, however, the wall is a sink of fluctuational energy, and the granular temperature decreases near the walls. The velocity profiles are shown in Fig 3.7(c) and (d). Both gas and solids velocities exhibit similar behaviors when e_w changes from 0.99 to 0.6 except that the jump of the gas velocity near the walls is more pronounced when e_w=0.99 due to the corresponding profiles of the granular temperature. Based on these results, it is clear that the flow structures are not sensitive to the different values of e_w. Further investigation is still required in order to explain the sensitivity to e_p.

In gas-particle flows, two kinds of coexisting forces are presented in the balance equations. One is the gas-particle interaction force, and the other is the interparticle interaction force. It is not clear which force causes the sensitivity of flow structures to e_p. It may be one force dominates this sensitivity or both of them are equally important. From a physical point of view, it is well known that the coefficient of restitution is dependent on the impact velocity as well as material properties (Lun and Savage, 1986). However, a theoretical model accounting for a velocity dependent e_p is very cumbersome (Lun and Savage, 1986). Thus normally the coefficient of restitution is considered as a constant in numerical calculations. A similar assumption is also applied in our model. Therefore, e_p is considered as a material property that only affects particle-particle interactions. So it is reasonable to assume that the sensitivity to e_p comes from the interparticle interaction force. To prove our hypothesis, a model describing granular flows in a channel is examined, based on the work of Wang et al. (1997). In this model, the particle flow is induced by gravity in the absence of a carrier gas. Using ρ_s, $(g\Delta)^{1/2}$ and Δ as characteristic density, velocity and length, the equations of motion for the fully-developed granular flow in a channel take the following form:

$$D\phi + \frac{d}{dy^*}\left(\frac{1}{2} f_3 (T^*)^{\frac{1}{2}} \frac{dv_x^*}{dy^*}\right) = 0 \tag{3.10}$$

$$\frac{d}{dy^*}(f_1 T^*) = 0 \tag{3.11}$$

$$\frac{d}{dy^*}\left(f_4 (T^*)^{\frac{1}{2}} \frac{dT^*}{dy^*}\right) - D^2 f_5 (T^*)^{\frac{3}{2}} + f_3 (T^*)^{\frac{1}{2}} \frac{1}{2}\left(\frac{dv_x^*}{dy^*}\right)^2 = 0 \tag{3.12}$$

Where $D = \Delta / d$, and dimensionless terms f_1 to f_5 are the same as those defined in Table 3.1. The boundary conditions for the solid velocity and pseudo thermal energy flux are still given by Equations 3.5 and 3.6, and the material properties and the bed width can be seen in Table 3.2. In this model, the positive vertical direction is downward. The effect of the coefficient of restitution (e_p) on the flow profiles of solids fraction, granular temperature and solids velocity is shown in Fig 3.8. It can be seen that the granular flow is still sensitive to e_p when its value is close to 1. When e_p is equal to 1, the wall is always a sink of the fluctuational energy. With a decrease in the value of e_p, the energy loss due to inelastic particle-particle collisions becomes pronounced and the flux of pseudo thermal energy towards the walls decreases, indicating a transition from sink wall conditions to source wall conditions. When e_p is smaller than 0.98, the granular flow is not sensitive to this parameter. By comparing Fig 3.5 and Fig 3.8, it can be seen that the detailed flow profiles of solids fraction and granular temperature are quite similar between gas-particle flows and granular flows except that the segregation of particles is a little more pronounced for gas-particle flows as e_p decreases. These results provide convincing evidence that the sensitivity to e_p essentially comes from the interparticle interaction terms, and the basic structures of the flow profiles for the solids fraction and granular temperature can be traced to granular flows.

To further investigate the origin of the sensitivity to e_p in granular flows in a channel, the profiles of stresses and pseudo-thermal energy across the channel are examined. In the kinetic theory closures, if the frictional contribution is neglected, the particle stress is composed of two terms, the kinetic stress and collisional stress (Lun et al., 1984). To the best of our knowledge, the distribution of kinetic stress and collisional stress across the bed has not been examined for continuum models for bounded gas-particle flows. Fig 3.9 shows the effect of the coefficient of restitution (e_p) on the distribution of solids phase stress through the channel. It is found that the magnitude of the total solids stress strongly depends on the value of coefficient of restitution. The increase in the total solid stress with an increase in e_p can be attributed to the fact that the system loses less energy as e_p increases, and the particles move faster and have a higher chance to collide with each other than in systems which are characterized by a lower coefficient of restitution. Similar results were reported by Liss and Glasser (2001) for a sheared granular flow. In Fig 3.9, it can be seen that the stress profile of the collisional contribution has a transition from a high level at the walls to a high level in the center when e_p is decreased from 1 to 0.95. This indicates that the distribution of particles should also have a transition with a decrease in e_p because the value of the collisional stress is proportional to the solids fraction. The transition of the kinetic contribution, however, is opposite with a decrease in e_p. Its maximum value switches from the center to the walls when e_p is decreased from 1 to 0.95. It is clear that large fluctuational velocity is expected to be seen in the region of high kinetic stress. Since the granular temperature is proportional to the mean square of fluctuational velocity, high kinetic stress should correspond to a large granular temperature and a small solids fraction. It is important to note that the transitions of the collisional stress and kinetic stress occur when e_p is decreased from 1 to 0.99. A further decrease in e_p from 0.99 to 0.95 just makes this transition increasingly more pronounced. The profiles of solid phase stress in gas-particle fluidized beds are also investigated (not shown here), and similar results are obtained to those seen in Fig 3.9. Therefore, it is clear that there is a relation between the sensitivity to e_p and the transition of the solids phase stress. However, the physical mechanism leading to this sensitivity is still not clear. Therefore, an analysis of the pseudo-thermal energy balance is required.

For the fully developed granular flow in a channel, the pseudo-thermal energy balance equation has three parts (see Equation 3.12). The first term represents the diffusive transport of PTE, E_f, the second term denotes the rate of dissipation of PTE through inelastic collisions, E_d, and the third term represents the rate of production of PTE by the work of particle-particle shear, E_w. The profiles of the contributions of these three parts with different values of e_p are shown in Fig 3.10. Since E_w is proportional to the gradient of the solids velocity, the symmetric boundary conditions lead to an E_w value of zero in the center. For elastic systems (Fig 3.10(a)) where the term E_d vanishes everywhere across the channel, E_f is zero at the center and keeps decreasing towards the walls. This means that the pseudo thermal energy has a maximum at the center and decreases towards the walls. It is well known that particles have a potential to segregate to the low pseudo thermal energy region. Thus the "Core-Annular" structure can be observed in elastic systems. When the collisions between particles are inelastic, E_d has a non-zero value at the center. To compensate this dissipation term, E_f must have a positive value. This indicates that the pseudo thermal energy has a low level at the center region and particles are forced to segregate there. This particle segregation towards the center occurs as soon as the value of e_p deviates from 1. Therefore, it is clear that the sensitivity to e_p is due to the positive value of E_d at the center region. To better understand the sensitivity of flow structures to the value of E_d, a factor α is multiplied to this dissipation term and its value is varied from 0 representing no energy loss from particle-particle collisions, to 1 representing the Lun constitutive relation for E_d. It is found that for a given value of e_p the flow profiles are very sensitive to the variation of this dissipation term. When α is equal to 0, a "Core Annular" flow is always obtained because of neglecting the effect of the energy dissipation. As the value of α increases to 0.1, however, a pronounced transition of the particle segregation from the walls to the center is observed. Thus, the sensitivity to e_p would be reduced with a reduction in the magnitude of the dissipation term, and a reduction in the magnitude of the dissipation term would also increase the range of the e_p values that lead to "Core-Annular" flow profiles.

3.2.2 Effect of Gravity

The second parameter examined is gravity (g). Fig 3.11 shows the variation of solids fraction amplitude, ϕ_{max}-ϕ_{min}, as a function of the gravity. The variation of the solids fraction amplitude with g is gradual and smooth, indicating that the flow structures are fairly insensitive to changes in this parameter. To further study the effect of g, the flow profiles of solids fraction, granular temperature, and gas and solid velocity are examined for g=9.81m/s^2 and g=1m/s^2 respectively. When the pressure drop is held constant, the gas and solid flow rates substantially increase with a decrease in gravity. However, a more appropriate comparison for a fluidized bed reactor is to keep the residence time or velocity of the gas a constant to complete reactions appropriately. Therefore, in the flow profile plots, we keep the average gas phase velocity a constant by reducing the pressure drop across the bed as gravity is decreased. The dimensional equations are used to analyze the effect of gravity in that the dimensionless model uses the solid terminal velocity as characteristic velocity and the solid terminal velocity is related to gravity. In Fig 3.12, elastic collisions between particles are assumed, so the "Core-Annular" flow structure is obtained. It can be seen that in this case the flow structures for constant average gas velocity are insensitive to the different values of gravity. The profiles of solids fraction and granular temperature are quite similar when g changes from 9.81 to 1 m/s^2, except that the particle segregation near the walls is slightly more pronounced when the gravity is lower. The gas and solid velocity profiles are also similar as g changes except that the solids velocity in the lower gravity case is slightly smaller.

The similarity of flow profiles with the variation of g is related to the compensation between the body force and the drag force, and the fact that we have kept the average gas phase velocity a constant. In a lower gravity system, although the body force, driving particles downward, is small, the drag force, keeping the particles in suspension, is also relatively small. To better illustrate this relation, the difference between the magnitude of the drag force and the magnitude of the gravity force is plotted in Fig 3.13. It can be seen that the difference between the drag force and the gravity force changes only slightly when the gravity changes nearly one order of magnitude. This result indicates that for a fluidized bed system with a high gas velocity the effect of changing the gravitational force can be automatically eliminated by the corresponding change of the drag force. Therefore, the base states are not sensitive to the parameter g. From Fig 3.12, it can be seen that particles always move upward under both high and low gravity situations when there is a high gas velocity. However, many riser flows operate with solids recirculating downward in the annular region close to the walls while still maintaining a positive net solids flux. Because the gravity causes the solids to flow downward, it is interesting to see the effect of the gravity on this flow regime.

Fig 3.14 shows that the effect of the gravity becomes significant as the average gas velocity changes from 10m/s (see Fig 3.12) to 1m/s. When the gravity is equal to 9.81m/s, the solids flow downward in the region close to the walls, although the average solids velocity is equal to 0.16m/s. When the gravity decreases to 1m/s^2, the average solids velocity increases to 0.7m/s and particles always move upward in the bed. So the particle annular downflow observed in the high gravity situation is no longer present. The difference between the magnitude of the drag force and the magnitude of the gravity force is plotted in Fig 3.15. In the center the value of the drag force minus gravity force is comparable between the high gravity case and the low gravity case. However, this value decreases rapidly under the high gravity situation when moving towards the walls, while this value only changes slightly under the low gravity situation. Since the drag force plays a key role in the study of the parameter g, an investigation of the effect of different expressions for the drag force was carried out.

Here, another correlation adopted by Wen and Yu (1966), Equation 3.13, was studied, and the flow profiles based on different gravity are shown in Fig 3.16.

$$\beta = \frac{3}{4} C_D \frac{\phi \rho_f \mid v - u \mid}{d(1-\phi)^{2.65}}$$

$$C_D = \frac{24}{R_{ep}} [1 + 0.15 (R_{ep})^{0.687} \qquad for \qquad R_{ep} < 1000$$

$$C_D = 0.44 \qquad for \qquad R_{ep} \geq 1000$$

$$R_{ep} = \frac{(1-\phi)\rho_f \mid v - u \mid d}{\mu_g} \tag{3.13}$$

It can be seen that with a high gas velocity, using Wen and Yu's (1966) constitutive relation for the drag force the gas-particle flows are still insensitive to the gravity except that the granular temperature becomes smaller with an increase in the gravity. In Fig 3.12 and Fig 3.16, the calculations for g=9.81m/s^2 were carried out under the same conditions except for different drag force expressions. A scrutiny of these two figures indicates that the flow profiles are not sensitive to the expressions of the drag force, which is in agreement with the results from Yasuna et al. (1995). When the average gas velocity is decreased, flow profiles similar to those in Fig 3.14 are obtained with Wen and Yu's constitutive relation for the drag force (not shown). The above observation could be useful for setting up fluidized bed reactors in sites with different gravity values, such as on the Moon and Mars, or centrifugal geometries. However, we would expect that the growth rate of instabilities would vary significantly with gravity. So unsteady state structures may be rather different (Gibilaro et al., 1986).

3.2.3 Effect of the Bed Width

So far we have considered gas-particle flows only in a narrow system, where the width of the bed is 4cm. In industry, the width (or diameter) of gas-fluidized beds is normally much larger than this. Therefore, it is of practical importance to extend our calculations to wider systems. The continuation of solids fraction amplitude, ϕ_{max}-ϕ_{min}, as a function of the dimensionless width D (Δ/d) is shown in Fig 3.17. It is clear that there are two branches in the continuation profile. In the first branch (A-B), initially the solids fraction amplitude increases with an increase in the bed width, after which the increase in the solids fraction amplitude slows down and a plateau curve is observed. As the dimensionless coefficient D increases to 350, the continuation profile switches to the second branch, in which the solids fraction amplitude linearly increases with an increase in the bed width. To better investigate the effect of the bed width on the flow structures, the flow profiles of solids fraction, granular temperature, and gas and solid velocity are plotted for Points A, B, C in Fig 3.18. For a narrow system, a "Core-Annular" flow structure is observed, given that elastic particle-particle collisions are assumed. However, with an increase in the bed width, the segregation of particles has a transition from the walls to the center. In branch A-B the solids fraction increases both at the walls and at the center with increasing D, and the maximum of the solids fraction is observed at the walls although the increase in the solids fraction is faster at the center than at the walls. When the value of D is above 250, the increase in solids fraction at the walls stops. Thus a plateau curve of solids fraction amplitude is observed. With continuously increasing the bed width, the maximum of solids fraction has a transition from the walls to center, and the curve switches to the B-C branch. The increase in the solids fraction at the center has a linear relation with an increase in the bed width. The magnitude of the granular temperature significantly increases with an increase in the bed width as a result of the growth of the gas and solid phase velocities when the gas pressure drop is held constant. It is found that the average gas phase velocity almost linearly increases with an increase in the bed width. Scrutiny of Fig 3.18(c) and Fig 3.18(d) reveals that in a narrow system the difference between the gas and solid velocity at the center is pronounced, while with an increase in the bed width this difference is reduced. This is because in Equation 3.2 the gas-particle relative velocity is multiplied by D, leading to a reduction in the relative velocity with an increase in D.

3.2.4 Effect of Mean Solids Fraction

The next parameter considered is the average solids fraction, based on the constraint of Equation 3.9. The average solids fraction is changed with the variation of the coefficient C_l as other physical parameters are held constants. In the present work, the effect of the average solids fraction is examined by plotting the continuation profile of the variation of the average solids fraction with C_l because it is not very meaningful to study the solids fraction amplitude if the average solids fraction in the system is different. For a narrow system (D=50), Fig 3.19 shows that the continuation of average solids fraction has three branches with the variations of C_l, while the upper branch indicates that the fluidized bed is almost operated in a packed bed state while the lower branch indicates that the fluidized bed is operated in a relatively low solids concentration state. The curve connecting these two branches represents a series of unstable solutions for the steady state. For a certain range of C_l, three solutions coexist in the system with the same set of physical parameters, while the lower one represents a dilute state, the upper one represents a packet bed state and the middle one denotes an unstable transition state. We find that if C_l is used as a parameter instead of the average solids fraction in the previous analysis, this "S" curve is a typical continuation profile for narrow gas-fluidized beds with the physical parameters used in the present work. As we track the variations of the average solids fraction with D and gravity respectively, similar continuation curves are obtained (see Fig 3.20).

When we extend our analysis to a wider system (D=500), two continuation branches coexist in the system. In Fig 3.21, initially the average solids fraction increases with an increase in C_l. As C_l passes through a maximum, the solution branch starts to turn back and the average solids fraction increases with a decrease in C_l. The continuation stops with the second branch, and the third branch observed in narrow systems representing the packet bed states is absent here. To better understand the variations of flow profiles with C_l, three points A, B, C are selected from Fig 3.21, and their solids fraction profiles are plotted in Fig 3.22. For a low solids fraction state (point A), particles are segregated near the walls and a flat curve is observed at the center. With an increase in average solids fraction, the particle segregation has a transition from the walls to the center. As the average solids fraction increases to 0.12 (point C), the particle concentration at the center is much higher than near the walls. Recalling the analysis of the effect of the variation of the bed width, it seems that an increase in the bed width or an increase in the average solids fraction enhance the migration of particles from the walls to the center. Therefore, when e_p is assumed equal to 1, a "Core-Annular" flow structure can be observed for a very dilute system with high bed width, while a core segregation of particles becomes pronounced for a wide system with a moderate solids load, where the average solids fraction is above 0.05. Reducing e_p just makes the core segregation more intensive. The transition from "Core Annular" structure to core segregation structure observed in the narrow system when the value of e_p is reduced from 1 cannot be seen in the wide system. Thus the sensitivity to e_p is reduced as the bed width increases.

The origin of the transitions observed in Fig 3.18 and 3.22 arises from the contribution of the gas-particle slip (J_{vis}) on the pseudo-thermal energy transport. By analyzing Equation 3.4, it is clear that the contribution of gas-particle slip increases with an increase in the solids fraction, and is greatly enhanced with an increase in the bed width because the dimensionless parameter D is multiplied by the damping contribution term and D^2 is multiplied by the production contribution term. If the gas-particle slip term is not considered in the model, particles always segregate near the walls when e_p=1 (see Fig 3.23) and the bed width and the average solids fraction have only a small effect on the transition of particle segregations.

3.2.5 Effect of Dimensionless Groups

Because the equations have been non-dimensionalized in the present work, it is interesting to look at the effects of dimensionless numbers. Generally, two dimensionless numbers are used to classify flow regimes and investigate instabilities in gas-particle fluidized beds. One has the form of Reynolds number Re, which is equal to $B*D*E$ in our model, and the other is the Froude number Fr, which is equal to $1/C$ in the calculations. It is important to note that here the dimensionless group Re is evaluated by the particle terminal velocity while the Reynolds number used to classify flow regimes in gas-fluidized beds is frequently evaluated by the gas phase velocity. From Figure 3.24, it can be seen that the steady state flows are not sensitive to Re. The flow profiles have no difference as Re changes from 2000 to 200 because the viscous term of the gas phase is relatively small in a high Re situation (see Equation 3.1) and the flow behavior is independent of Re without this term. With a decrease in Re, the gas viscous term becomes important in Equation 3.1, and gas and particle velocities are decreased. The influence of the Froude number is more pronounced (see Fig 3.25). When the average solids fraction and gas pressure drop are held constant, the gas and particle phase velocity decrease with an increase in the Froude number, leading to a decrease in the granular temperature. When this dimensionless number is high enough, the contribution of the viscous terms for both gas and particle phases can be neglected, and the system can be simplified to a force balance between the gas-particle drag force and the bed gravity force. Since a higher granular temperature means a higher particle pressure and particle viscosity, which can suppress instabilities in gas-particle fluidized beds, a gas-fluidized bed with a lower Froude number is likely to be more stable, which is in agreement with the analysis of Wilhelm and Kwauk (1948).

3.3 Conclusions

In this chapter, a laminar continuum model is presented to describe gas-particle flows in a channel with mono-disperse particles. The steady state solutions obtained by a numerical method based on orthogonal collocation approach, exhibit a variety of structures. The effects of physical parameters on the steady state flow profiles are examined, using a continuation approach.

Flow properties are sensitive to the coefficient of particle-particle restitution when its value is close to 1. The particle concentration has a transition from segregation near the walls to segregation at the center when e_p decreases from 1 to 0.99. This sensitivity is reduced when the system becomes more inelastic. Since e_p is considered a material property, this sensitivity essentially comes from the particle-particle interaction terms. By examining a model describing a granular flow in a channel driven by gravity, a transition in the profiles of particle phase stress is observed. The kinetic contribution is small in regions where the collisional contribution is large, and vice versa. In elastic systems, the kinetic stress has a high level at the center and a low level near the walls. As the system changes from an elastic condition to a slightly inelastic condition, the high kinetic stress switches from the center to the walls. The physical mechanism leading to the origin of the sensitivity to e_p is studied by an investigation of the PTE profiles across the channel. It is clear that a non-zero value of the dissipation energy loss in the center region leads to this sensitivity.

It is useful to examine the sensitivity of these results to different closures to solids-phase stress. One model we have examined was proposed by Sela and Goldhirsch (1998), who used a more accurate calculation approach to obtain solid phase constitutive relations. In this model, the Boltzmann equation was perturbatively solved, and the constitutive relations were derived by performing a double expansion of the single-particle distribution in the Knudsen number and the degree of inelasticity. In Lun's model, instead, an ansatz for the form of the single-particle distribution was used to substitute into the Enskog equations. We have solved gas-particle flows in a channel using the closures of Sela and Goldhirsch (1998). The particle segregation still has a transition from the walls to the center when e_p decreases from 1 to 0.99 (see Fig 3.26).

Another question that arises is the effect of using an impact velocity dependent e_p to replace the constant e_p used in the present simulation. However, the closures incorporating an impact velocity dependent e_p are not easily obtained. To the best of our knowledge, only Lun et al. (1986) reported a simple estimation for these terms. In their approximation, the impact velocity is proportional to the square of the granular temperature. Since for slightly inelastic systems only the energy dissipation term in Equation 3.4 is sensitive to the variation of e_p, a modified model is established by changing the energy dissipation term based on an impact velocity dependent e_p, and holding other terms unchanged.

$$J_{coll} = \frac{12\rho_s\phi^2 g_0}{\pi^{1/2} d} C^* T^{3/2}$$

$$C^* = -4\overline{T} + (\pi\overline{T})^{1/2}(3 + 8\overline{T})\exp(4\overline{T})erfc(2\overline{T}^{1/2}) \tag{3.14}$$

where *erfc(x)* is the usual complementary error function which is approximated using rational functions (Abramowitz and Stegun, 1965), and the dimensionless $\overline{T}$ is equal to $C*T$ and the value of C is about (0.005^2) according to experimental measurements. Fig 3.27 shows that using this modified model, the particles always segregate in the center and the "Core-Annular" flow structure is not observed.

The second parameter we consider in this paper is gravity. It is found that the variation of the solids fraction amplitude with g is very smooth as g changes more than one order of magnitude. When the pressure drop is held constant, the gas and solid flow rates increase considerably with a decrease in gravity. When the average gas velocity is held constant, the flow profiles are not sensitive to this parameter when the gas velocity is high, which is attributed to the automatic compensation of the gas-particle drag force to the gravitational force. When the gas flow rate is low, however, particles move downward in the region close to the walls with high gravity, while this downflow is not observed under a low gravity situation.

In the simulations, it is found that the transition in the solids fraction profiles is not only dependent on the parameter e_p. If e_p is assumed equal to 1, the segregation of particles has a potential to transit from the walls to the center if the gas-particle system changes from a narrow fluidized bed to a wide fluidized bed, or from a dilute fluidized bed to a dense fluidized bed. This transition is explained by the contribution of the gas-particle slip on the pseudo-thermal energy transport. If this term is eliminated, a "Core Annular" flow structure is observed as long as the collisions between particles are elastic. In a narrow system, the continuation shows a "S" form bifurcation diagram when we track the variation of the average solids fraction with a certain parameter. However, when we extend our study to wider systems, the upper branch observed in narrow systems representing the packed bed state is absent. Therefore, the continuation diagram has a "U" form. The gas phase velocity, solid phase velocity and pseudo-thermal temperature greatly increase with an increase in the bed width.

The effect of two dimensionless numbers has been also studied in the present work. It is found that the flow behavior is not sensitive to the variation of the (terminal velocity based) Reynolds number when this number is high because, under this situation, the contribution of the gas viscous term on the gas phase momentum balance can be neglected. The effect of the Froude number is more pronounced in our physical situation. The granular temperature decreases with an increase in the Froude number, as well as the gas and particle velocities.

Based on the above analysis, it is found that the "Core-Annular" flow structure can be obtained only under certain conditions, using this kinetic theory model. In most cases, the simulations admit solutions representing a form of core segregation particles. However, in most fast gas-fluidized bed experiments, the "Core-Annular" flow structure was observed. To solve this conflict, two possible hypotheses are proposed here. The first is that in most experiments the "Core-Annular" flows observed were not steady state structures. Therefore, the comparison between the experimental measurements and the steady state simulations is not very meaningful. The second hypothesis is that the solid core segregation profile obtained in the simulation is unstable to small perturbations. Thus, this structure is not observed in the fluidized bed experiments. A goal of work in this area is a full bifurcation and stability analysis (of steady and time dependent solutions) and the results presented in this chapter represent a first step in this regard. A linear stability analysis of the steady state solutions is shown in the next chapter.

Table 3.1: Dimensionless functions f_0-f_8

$$f_0 = \phi(1-\phi)^{1-N}$$

$$f_1 = \phi(1+4\eta\phi g_0)$$

$$f_2 = \frac{81\mu_{eg}^2}{g_0\sqrt{\pi}}\phi$$

$$f_3 = \frac{2+\alpha}{3}\left[\frac{5\sqrt{\pi}}{48g_0\eta(2-\eta)}(1+\frac{8}{5}\eta\phi g_0)(1+\frac{8}{5}\eta(3\eta-2)\phi g_0)+\frac{16}{5}\eta\frac{\phi^2 g_0}{\sqrt{\pi}}\right]$$

$$f_4 = \frac{25\sqrt{\pi}}{16\eta(41-33\eta)g_0}\left[(1+\frac{12}{5}\eta\phi g_0)(1+\frac{12}{5}\eta^2(4\eta-3)\phi g_0)+\frac{64}{25\pi}(41-33\eta)(\eta\phi g_0)^2\right]$$

$$f_{4h} = \frac{25\sqrt{\pi}}{16\eta(41-33\eta)g_0}(\frac{1}{\phi}+\frac{12}{5}\eta g_0)\frac{12}{5}\eta(2\eta-1)(\eta-1)\frac{d}{d\phi}(\phi^2 g_0)$$

$$f_5 = \frac{48}{\sqrt{\pi}}\eta(1-\eta)\phi^2 g_0$$

$$f_6 = \frac{\sqrt{3}}{3}\frac{\pi\phi g_0}{\phi_{\max} f_3}$$

$$f_7 = \frac{\sqrt{3}}{6}\frac{\pi\phi g_0}{\phi_{\max} f_4}$$

$$f_8 = \frac{\sqrt{3}}{4}\frac{\pi\phi g_0}{\phi_{\max} f_4}$$

Table 3.2: Basic simulation parameters

d	Particle diameter	400μm
μ_g	Viscosity of gas	0.00004 kg/(ms)
ϕ_{max}	Maximum solid volume fraction	0.65
ρ_g	Gas density	3 kg/m^3
ρ_s	Solid density	1600 kg/m^3
N	Richardson-Zaki number	3
g	Gravitational acceleration	9.81 N/kg
2Δ	Bed width	4cm
e_w	Particle-wall restitution coefficient	0.9
φ'	Specularity coefficient	0.5

Fig 3.1: Effects of gas boundary conditions on the flow profiles. ($e_p=1$, $\bar{\phi}=0.1$) a) gas velocity profile b) solids fraction profile.

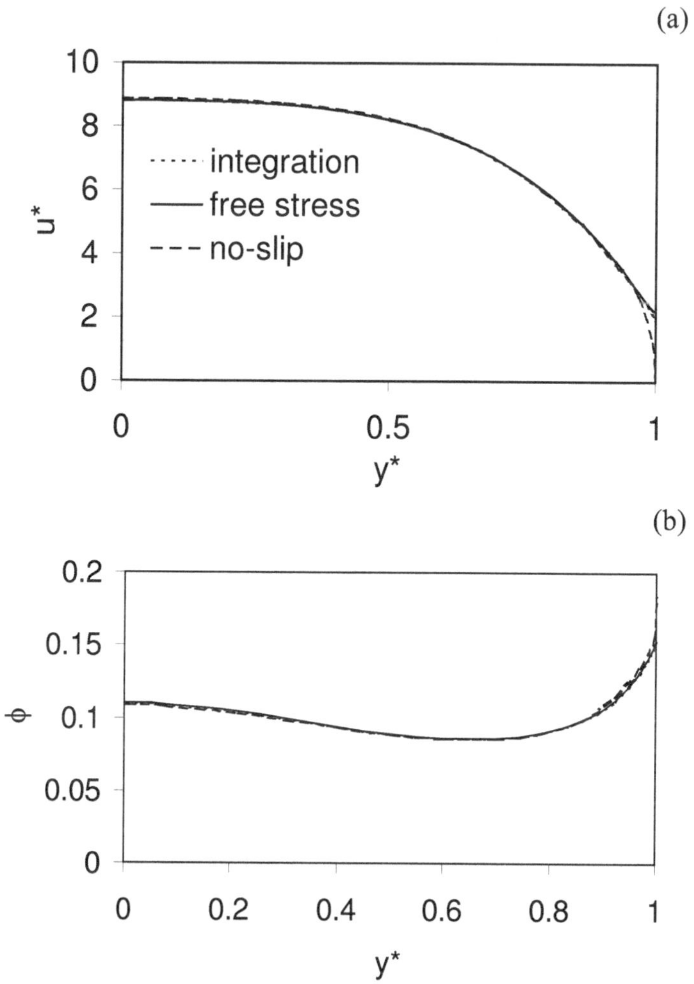

Fig 3.2: Gas flow rate vs. solid flow rate at a low pressure gradient. (dp^*/dx^*=-0.4 e_p=1)

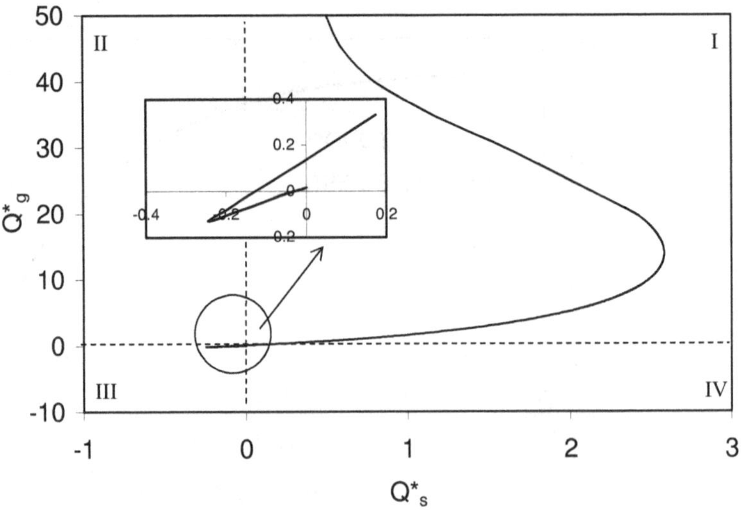

Fig 3.3: Gas flow rate vs. solid flow rate at a high pressure gradient. ($dp^*/dx^*=-1$ $e_p=1$)

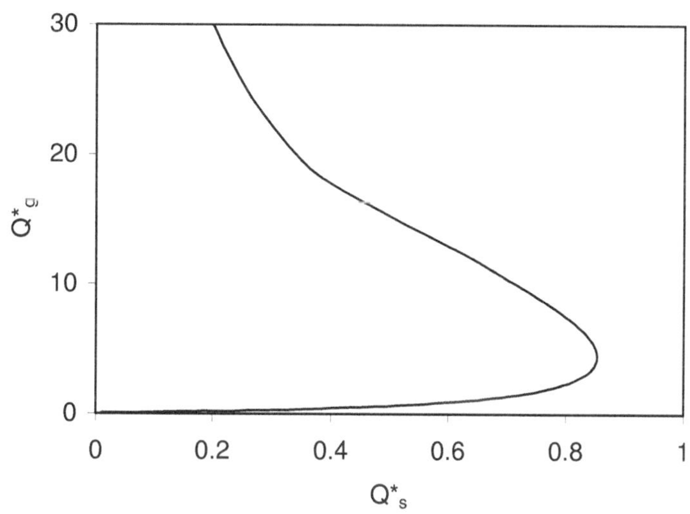

Fig 3.4: Variation of the solids fraction amplitude, ϕ_{max}-ϕ_{min}, with e_p. ($\bar{\phi}$=0.2)

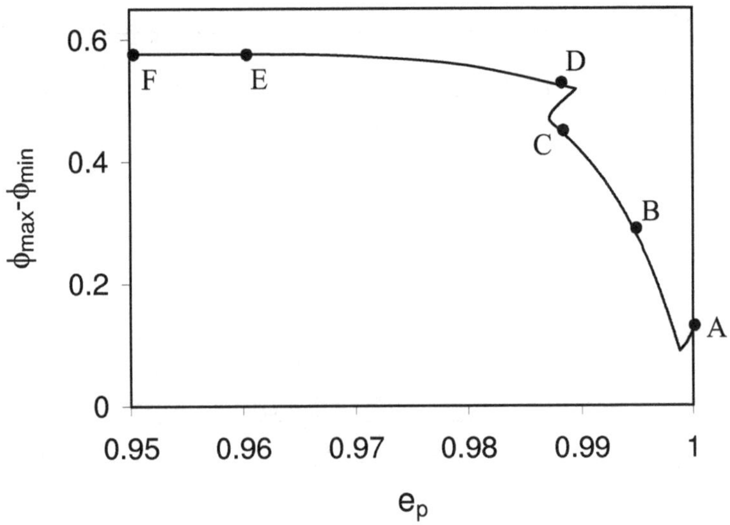

Fig 3.5: Effects of the coefficient of particle-particle restitution on the gas-particle channel flow. ($\bar{\phi}$=0.2) a) solids fraction profile b) granular temperature profile c) gas velocity profile d) solid velocity profile. (Points A, B, E and F are given in Fig 3.4)

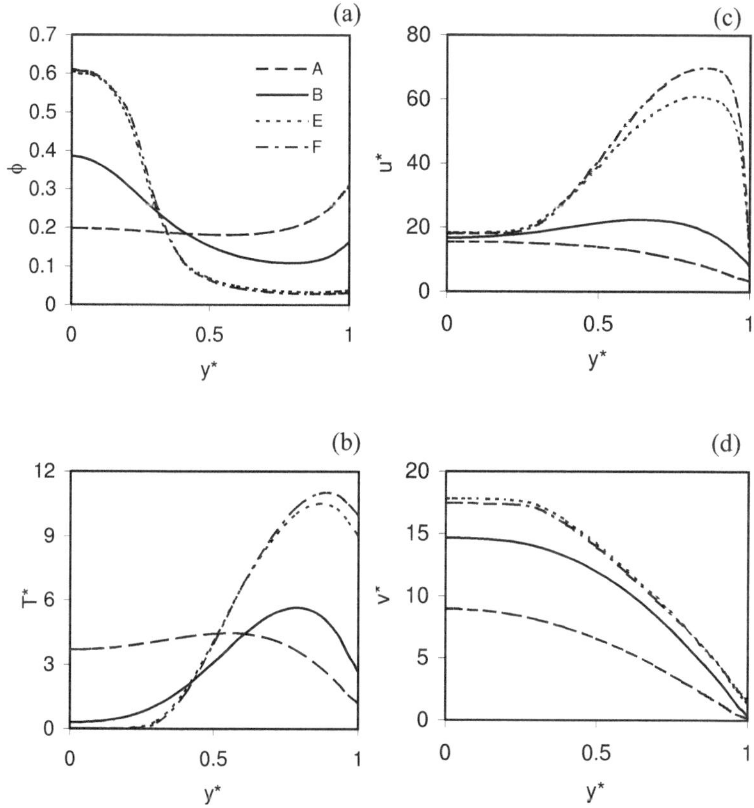

Fig 3.6: Solids fraction profiles corresponding to Points C and D in Fig 3.4. ($\bar{\phi}$=0.2)

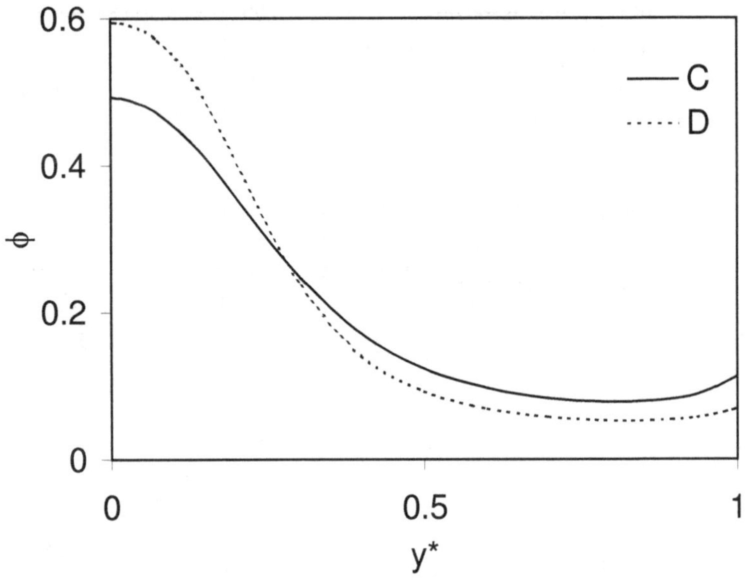

Fig 3.7: Effects of the particle-wall restitution coefficients on the gas-particle channel flow. ($\bar{\phi}$=0.25, e_p=0.96) a) solids fraction profile b) granular temperature profile c) gas velocity profile d) solid velocity profile.

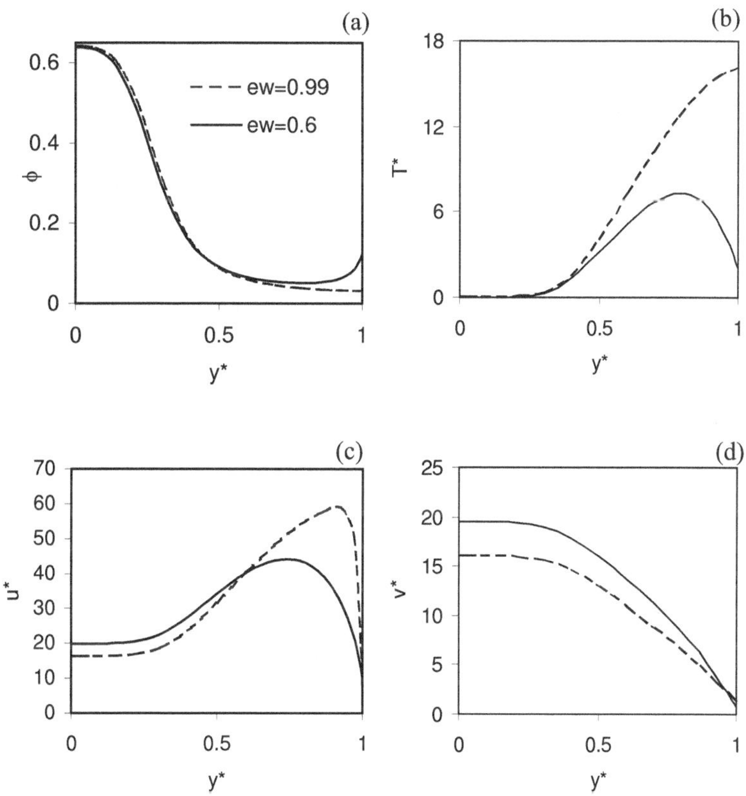

Fig 3.8: Effects of the coefficient of particle-particle restitution on the granular channel flow. ($\bar{\phi}$=0.2) a) solids fraction profile b) granular temperature profile c)solid velocity profile.

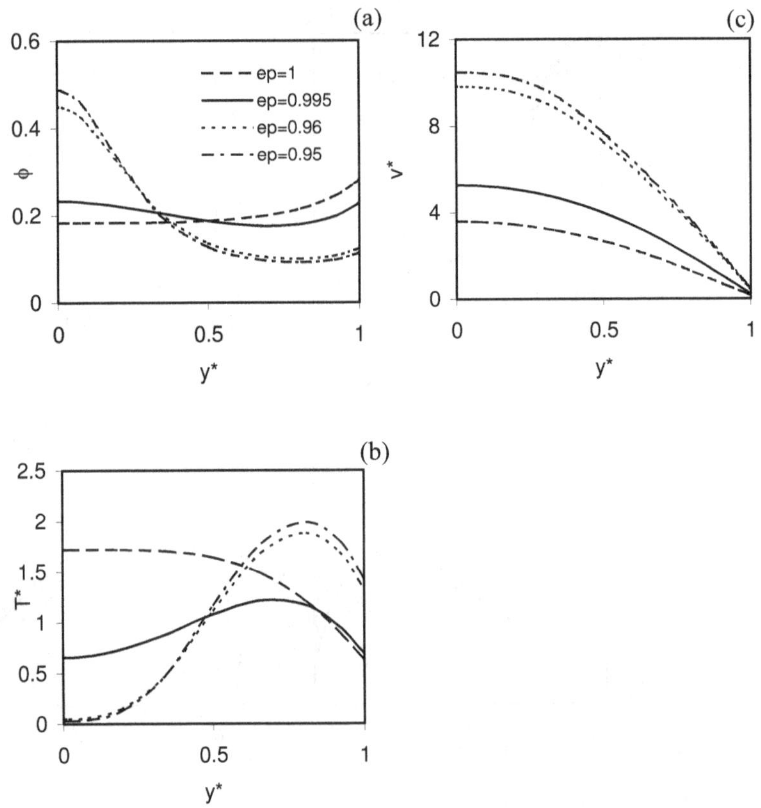

Fig 3.9: Particle stress profiles in the granular channel flow. ($\bar{\phi}$=0.1) a) e_p=1 b) e_p=0.99 c) e_p=0.95.

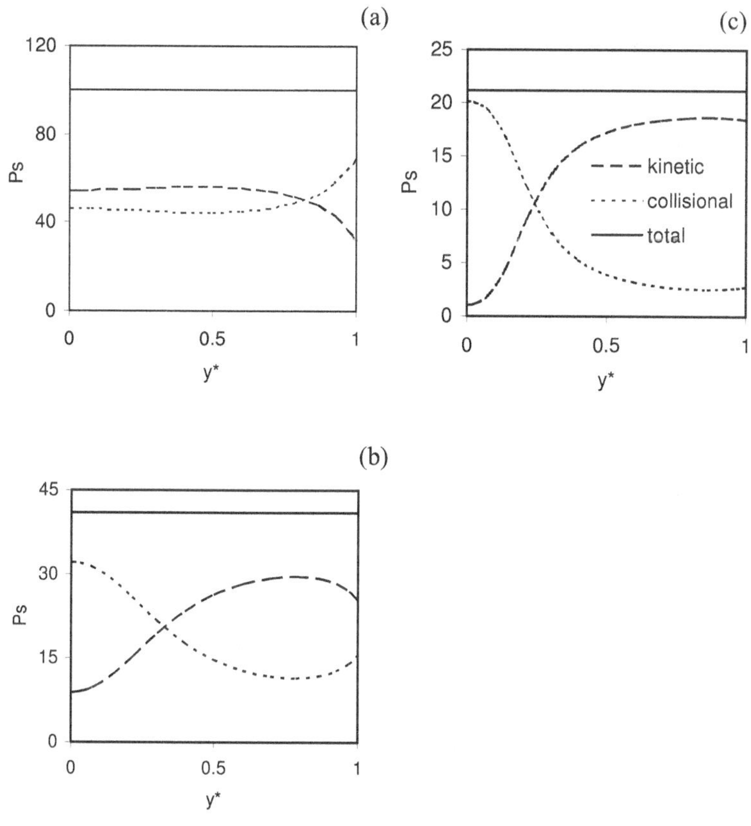

Fig 3.10: Pseudo-thermal energy profiles in the granular channel flow. ($\bar{\phi}$=0.1) a) e_p=1 b) e_p=0.99 c) e_p=0.95.

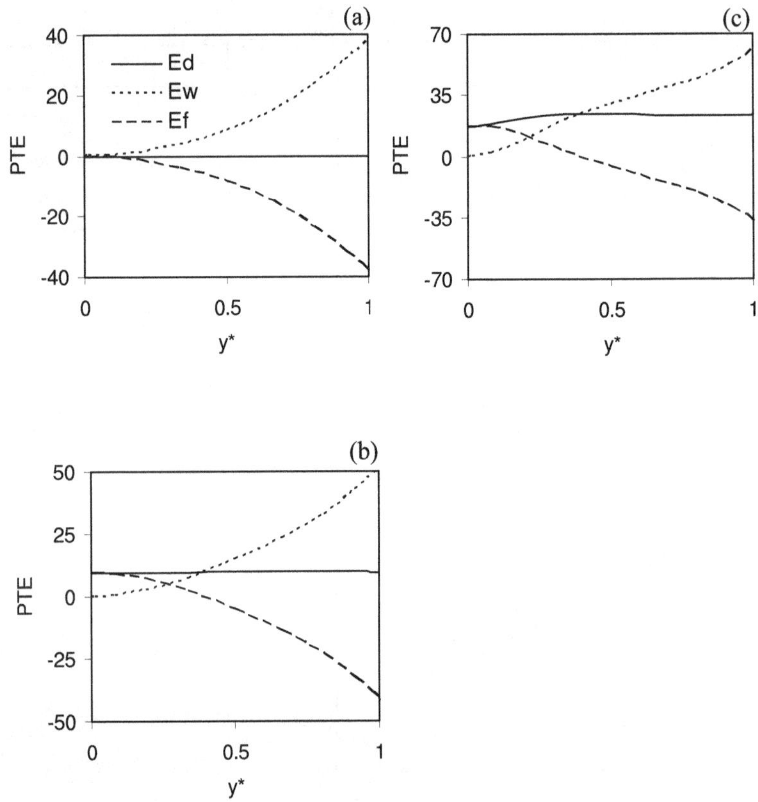

Fig 3.11: Variation of the solids fraction amplitude, ϕ_{max}-ϕ_{min}, with gravity. ($\bar{\phi}$=0.1, e_p=1)

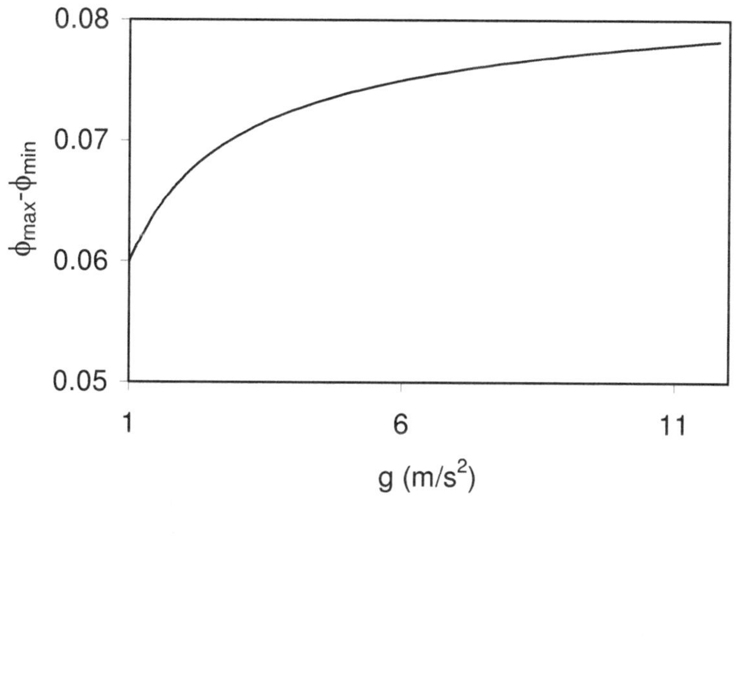

Fig 3.12: Effects of gravity on the gas-particle channel flow with a high average gas velocity. (u_g=10m/s, e_p=1, $\bar{\phi}$=0.12) a) solids fraction profile b) granular temperature profile c) gas velocity profile d) solid velocity profile.

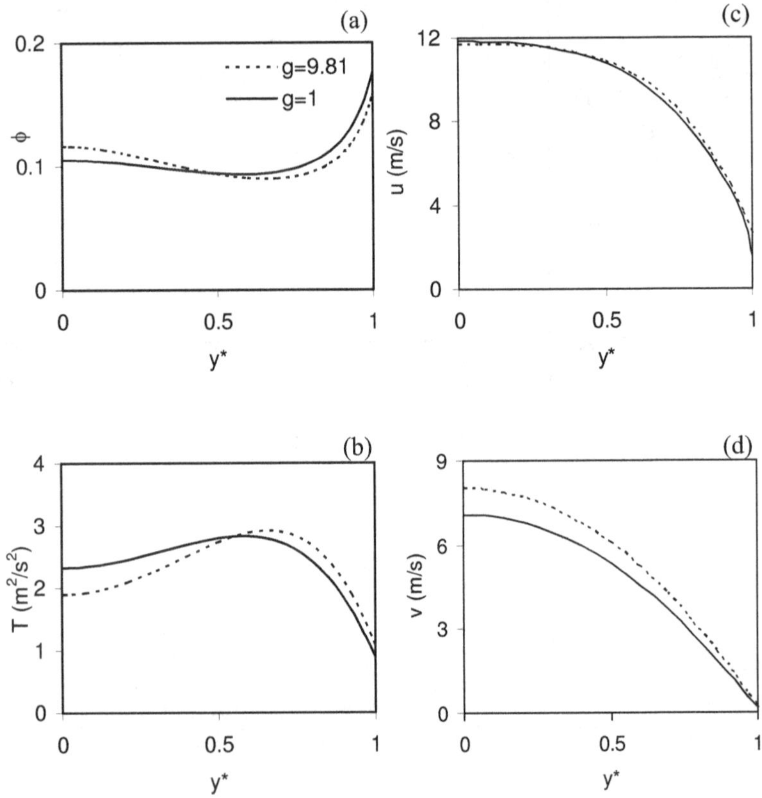

Fig 3.13: Profiles of drag force minus gravity force across the bed with different gravity.

(u_g=10m/s, e_p=1, $\bar{\phi}$=0.12)

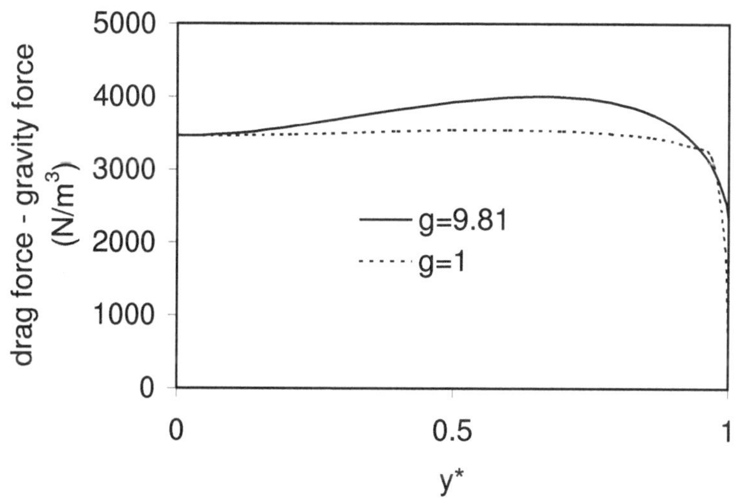

Fig 3.14: Effect of gravity on the gas-particle channel flow with a low average gas velocity. (u_g=1m/s, e_p=1, $\bar{\phi}$=0.22) a) solids fraction profile b) granular temperature profile c) gas velocity profile d) solids velocity profile.

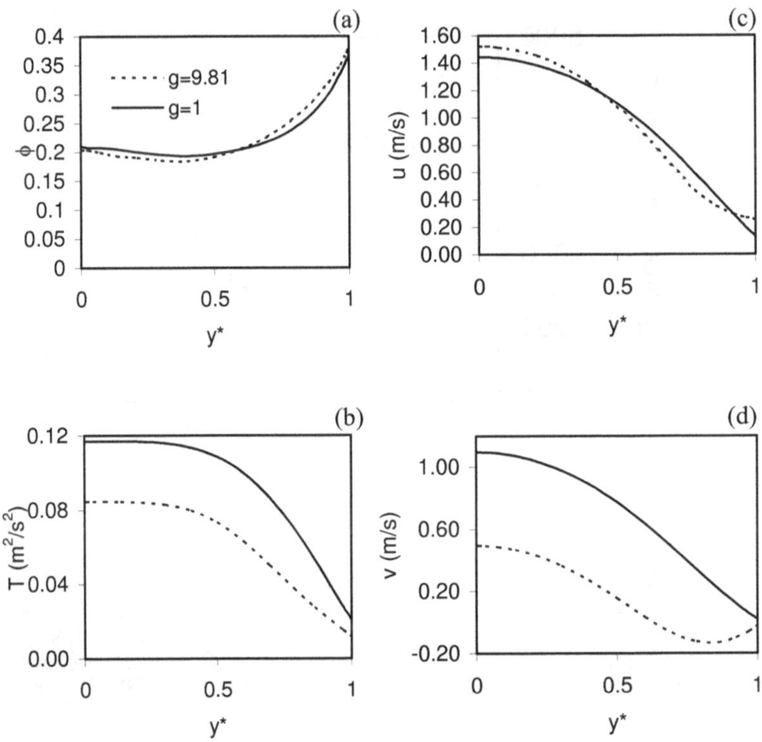

Fig 3.15: Profiles of drag force minus gravity force across the bed with different gravity.

(u_g=1m/s, e_p=1, $\bar{\phi}$=0.22)

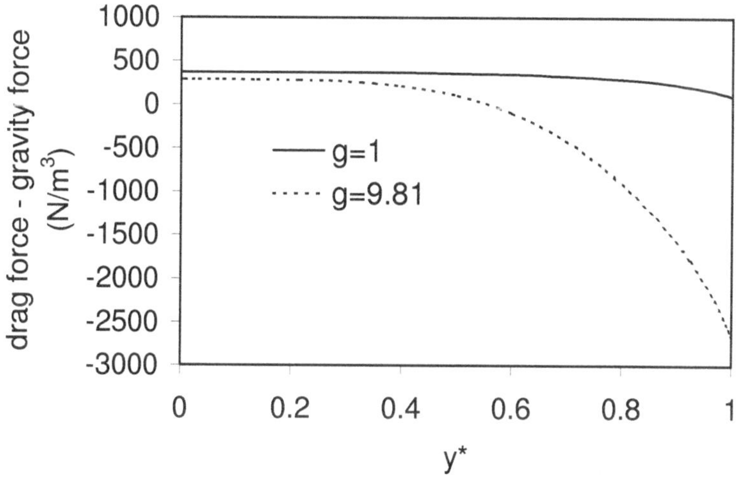

Fig 3.16: Effect of gravity on gas-particle channel flows with constant gas velocity, using Wen and Yu's (1966) correlation for the drag coefficient. (u_g=8.5m/s, e_p=1, $\bar{\phi}$=0.12) a) solids fraction profile b) granular temperature profile c) gas velocity profile d) solid velocity profile.

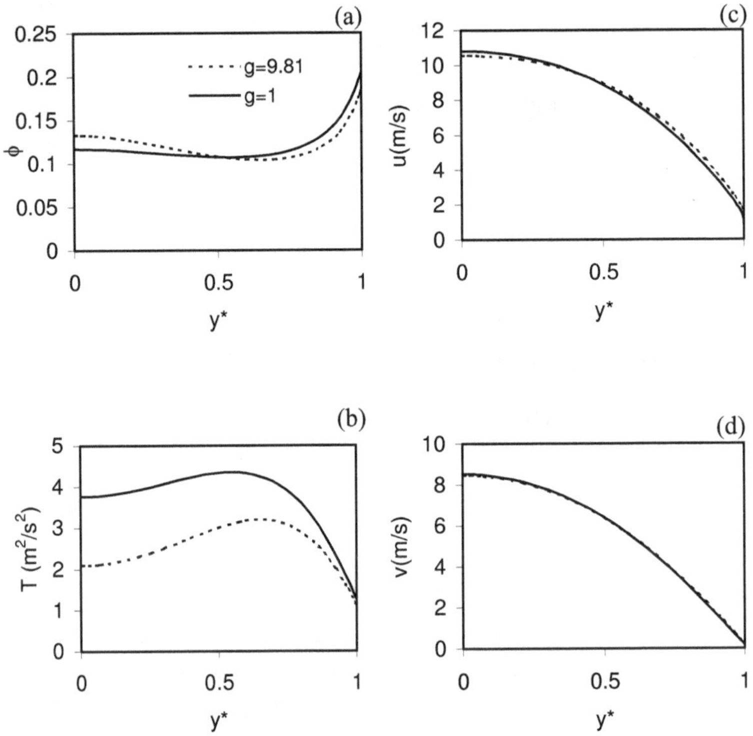

Fig 3.17: Variation of the solids fraction amplitude, ϕ_{max}-ϕ_{min}, with the dimensionless coefficient D (Δ/d). ($\bar{\phi}$=0.1, e_p=1)

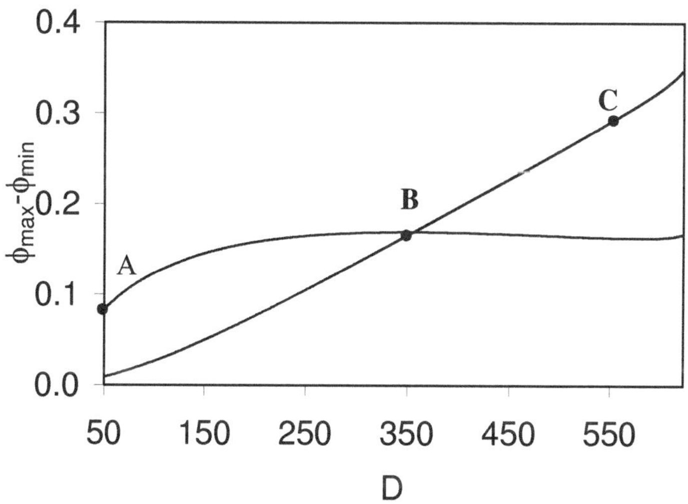

Fig 3.18: Effects of the bed width on the gas-particle channel flow. ($\bar{\phi}$=0.1, e_p=1) a) solids fraction profile b) granular temperature profile c) gas velocity profile d) solid velocity profile. (Points A, B and C are given in Fig 3.17)

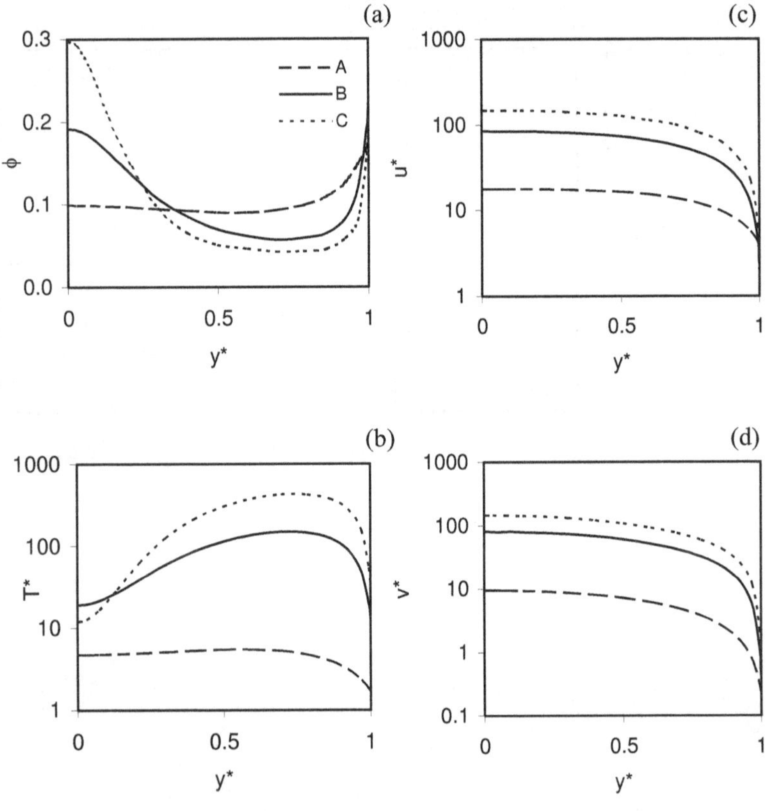

Fig 3.19: Variations of the average solids fraction with the coefficient C_1 in a narrow system. (e_p=1, D=50)

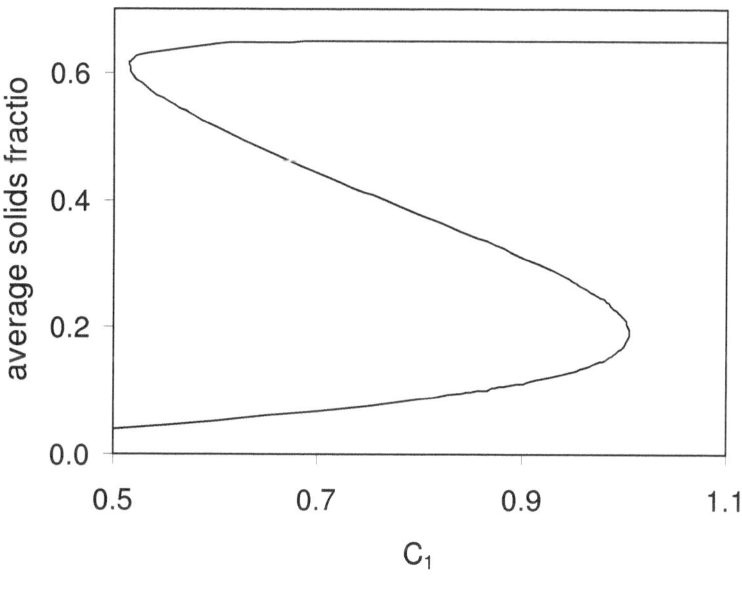

Fig 3.20: Variations of the average solids fraction with physical parameters. (e_p=1, D=50). a) the effect of the dimensionless coefficient D (Δ/d) b) the effect of gravity

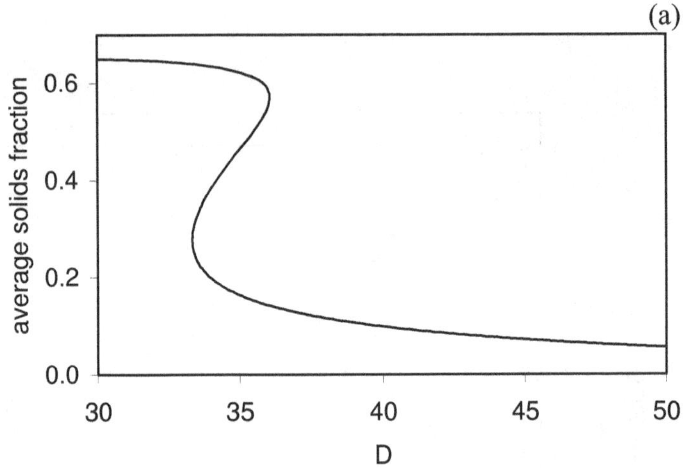

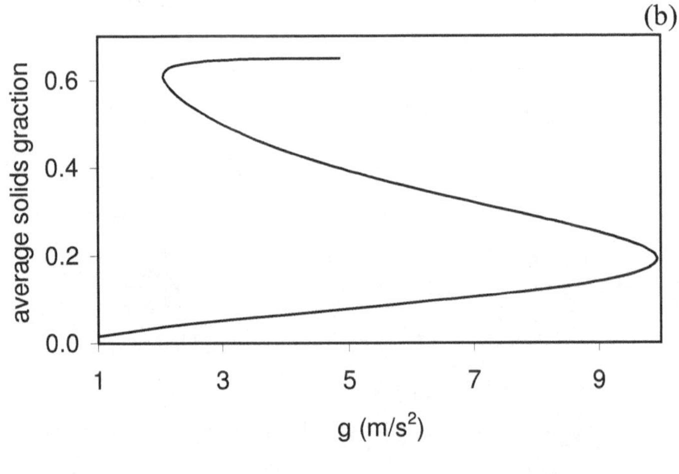

Fig 3.21: Variations of the average solids fraction with the coefficient C_1 in a wide system. (e_p=1, D=500)

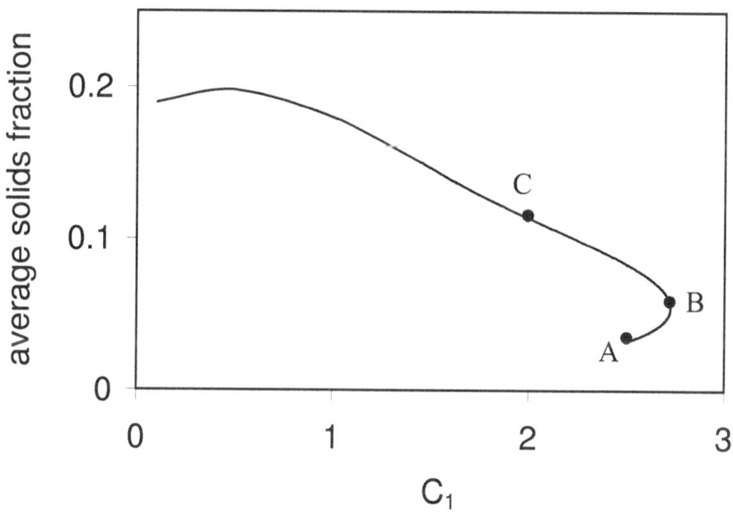

Fig 3.22: Solids fraction profiles corresponding to Points A, B and C in Fig 3.21. (e_p=1, D=500)

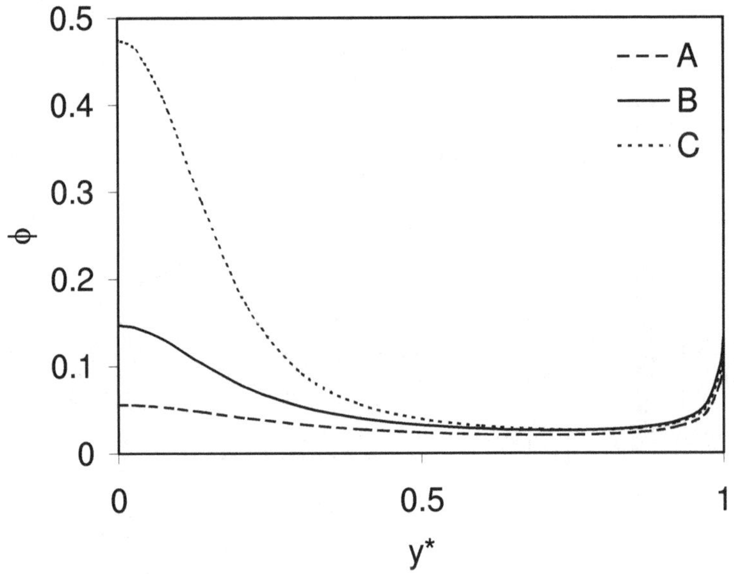

Fig 3.23: Solids fraction profiles for different average solids fractions when the contribution of gas-particle slip term (J_{vis}) is not considered. (e_p=1, D=500)

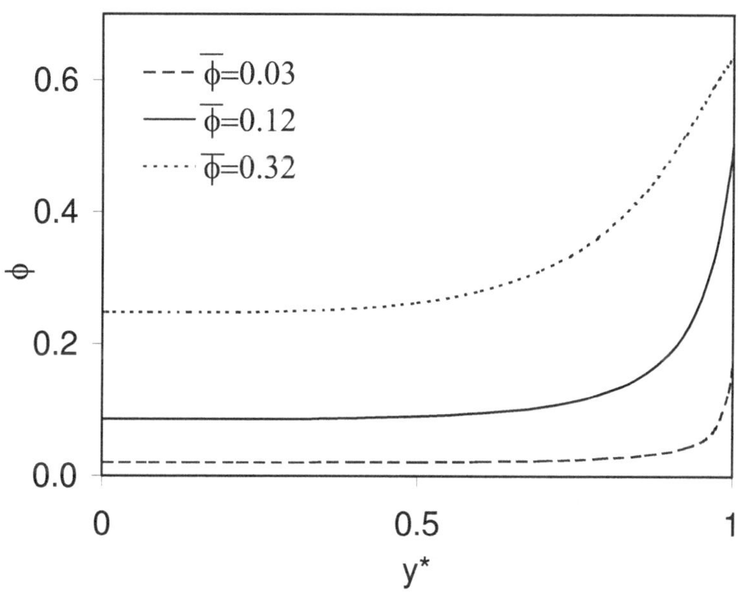

Fig 3.24: Effects of Reynolds number on the gas-particle channel flow. ($\bar{\phi}$=0.1, e_p=1) a) solids fraction profile b) granular temperature profile c) gas velocity profile d) solid velocity profile.

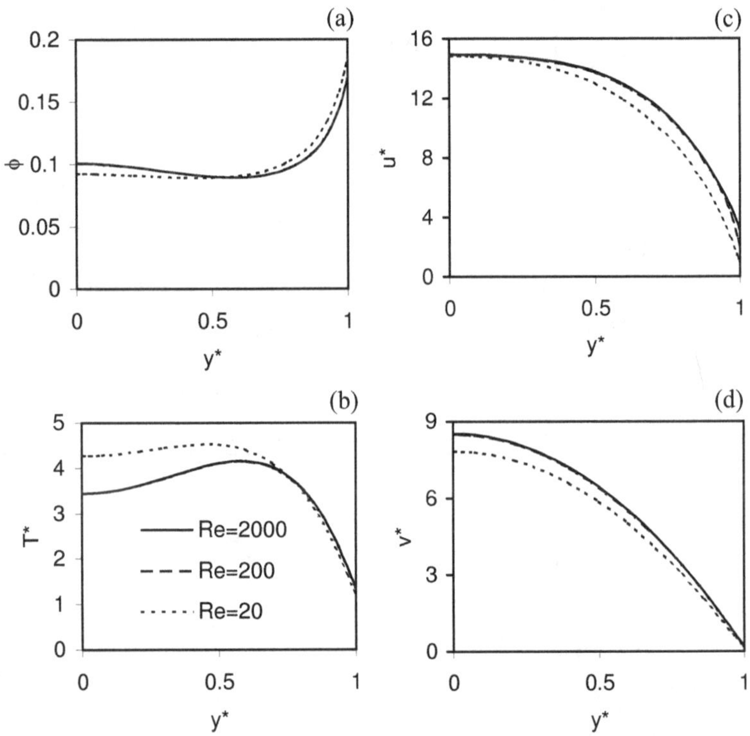

Fig 3.25: Effects of Froude number on the gas-particle channel flow. ($\bar{\phi}$=0.1, e_p=1) a) solids fraction profile b) granular temperature profile c) gas velocity profile d) solid velocity profile.

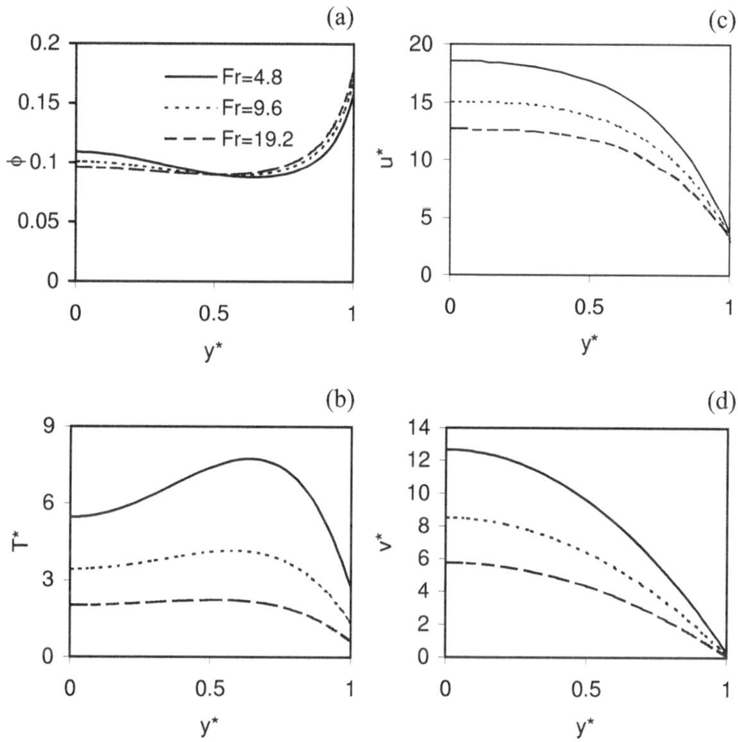

Fig 3.26: Flow profiles based on the Sela and Goldhirsch (1998) model. ($\bar{\phi}$=0.2, D=50)

a) solids fraction profile b) granular temperature profile.

(a)

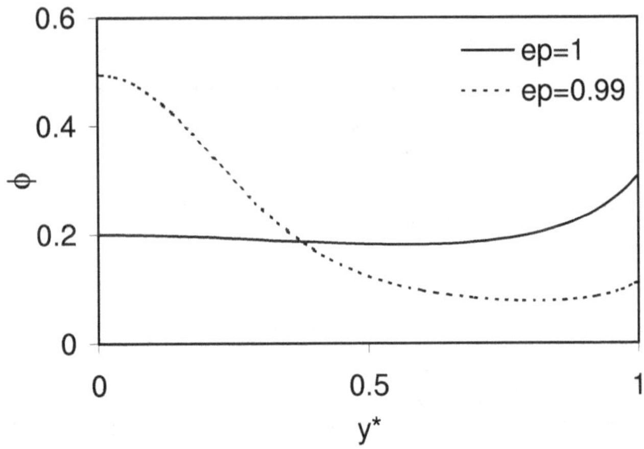

(b)

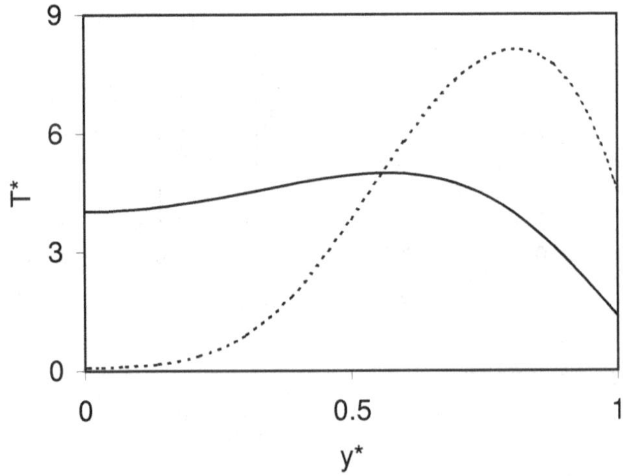

Fig 3.27: Effect of velocity dependent e_p on the flow profiles. ($\bar{\phi}$=0.2, D=50) a) solids fraction profile b) granular temperature profile.

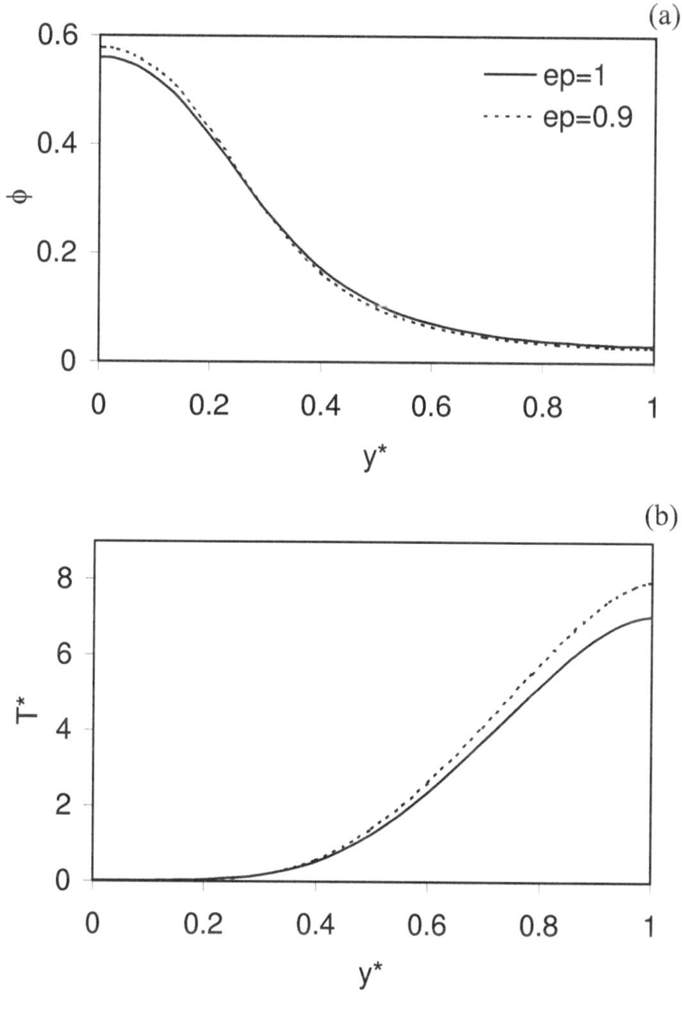

Chapter 4 Instability of Gas-Particle Fluidized Beds with Monodisperse Particles

In the previous chapter, the steady state solutions for bounded gas-particle flows were computed. However it is not clear whether these base state flow profiles can be observed in real fluidized bed operations. Therefore, a stability analysis on these base state solutions is required. In this chapter we would like to answer the following questions. Are the steady state solutions of this continuum model stable? What is the nature of instabilities? How do the physical parameters affect the instabilities? How can we predict the generation of voidage waves, such as bubbles, clusters and streamers? How do the flow hydrodynamics affect the chemical reactions? By answering these questions, some important features associated with the fluctuation behaviors in gas-particle fluidized beds can be explained. For example, if the steady state solutions are always stable then we know that this kinetic theory model is missing some important physics because time-dependent flows can be observed experimentally. On the other hand, if the steady state solutions are always unstable, then the compare between the steady state simulations and the experimental measurements is not very meaningful. Instead, the time-dependent simulations should be investigated.

4.1 Linear Stability Analysis for Unbounded Gas-Particle Fluidized Beds

To study the effect of walls, the unbounded flows should be investigated in advance.

4.1.1 Steady State Solutions and Numerical Method

Without boundaries, the equations of motion have a solution representing a uniformly fluidized bed with

$$\phi = \phi_0, \qquad \underline{v} = 0, \qquad \underline{u} = \underline{j}u_0, \qquad T = T_0 \tag{4.1}$$

where $\underline{j}$ is a unit vector pointing vertically upward, and ϕ_0, u_0 and T_0 are constants. The value of u_0 and T_0 can be obtained from Equation 4.2 and 4.3 respectively.

$$\beta u_0 = \phi_0(\rho_s - \rho_f)g \tag{4.2}$$

$$J_{coll} + J_{vis} = 0 \tag{4.3}$$

Then the base state solution is augmented by small perturbations, and the equations of continuity, momentum and pseudo-thermal energy are expanded in ascending powers of the perturbations through a Taylor's expansion of nonlinear terms. Since the disturbances are assumed to be both small and smoothly varying in space and time, their derivatives are also small. Therefore, all terms of degree greater than one in the perturbations are then neglected, leading to a set of linear partial differential equations in the perturbations. In the linear stability analysis approach, the solutions for the perturbations are assumed to have the form of plane waves. For example, the perturbation of the solids fraction can be written as

$$\phi_1 = \hat{\phi}_1 \exp(st)\exp(\underline{K} \cdot \underline{X}) \tag{4.4}$$

where ϕ_1 is the perturbation associated with the solids fraction, and $\hat{\phi}_1$ is a constant representing the amplitude of ϕ_1. $\underline{K}$ represents the wave number vector of the plane wave disturbance, whose wavelength λ_i in the i direction (i=x, y, z) is therefore given by λ_i $=2\pi/|K_i|$. For a two-dimensional simulation, $exp(\underline{K}\cdot\underline{X})$ can be rewritten as

$exp(k_x*x)exp(k_y*y)$, where k_x represents the wave number in the vertical direction while k_y denotes the one in the lateral direction. In general s is a complex number which may be written as $s=\xi-i\omega$. The imaginary part ω then determines v_p, the velocity of propagation of the waves, according to $v_{pi}=\omega/|K_i|$, while the real part ξ, defined as the growth rate, determines the rate of growth or decay of the waves with time. If ξ is positive, the disturbance grows and the base state is unstable, while if ξ is negative, the disturbance decays and the uniform state is called stable. The growth distance, Δx, can be defined by $\Delta x_i=\omega/\xi|K_i|$.

4.1.2 Instability in Unbounded Gas-Particle Fluidized Beds

Fig 4.1 shows the result of the growth rate verse the vertical wave number when k_y is zero, indicating that the simulation represents a one-dimensional stability analysis to the uniformly fluidized state with vertical disturbances. Under our specific situation, the gas-particle fluidized bed system is unstable in a certain range of vertical wave numbers and an increase in the average solids fraction can reduce the instabilities. We believe this is due to the variation of the particle pressure and the particle viscosity with the solids fraction because it is well known that the particle pressure and the particle stress can suppress the growth of disturbances. Garg and Prichett (1975) found that the uniformly fluidized bed could be stabilized if the particle pressure increases sufficiently rapidly with the increase in the particle concentration. Therefore, to better understand the physical mechanism of the stability behaviors in the unbounded system, the effects of the solids fraction on the particle pressure and the effective particle viscosity are investigated.

Using the Lun's model, the constitutive relations of the particle pressure and the particle viscosity take the forms,

$$\underline{\underline{P_s}} = (\rho_s \phi T + \rho_s 4\eta\phi^2 g_0 T)\underline{\underline{I}} \tag{4.5}$$

$$\mu_s = \frac{2\mu}{\eta(2-\eta)g_0}[1+\frac{8}{5}\eta(3\eta-2)\phi g_0] + \frac{16\mu\phi}{5(2-\eta)}[1+\frac{8}{5}\eta(3\eta-2)\phi g_0] \tag{4.6}$$

The variations of the particle pressure with the solids fraction can be seen in Fig 4.2(a). With an increase in the solids fraction from zero, the particle pressure increases, passes through a maximum and then decreases. As the solids fraction approaches the close-packed limit, the particle pressure increases again. Finally, a finite value, around 0.27Pa is obtained at ϕ_{max}. This variation of the particle pressure with the solids fraction were also observed by Zenit et al. (1997) and Koch and Sangani (1999). Fig 4.2(b) shows that the effective particle viscosity increases monotonically with an increase in the solids fraction. A sharp increase can be observed as the solids fraction is larger than 0.6. When the value of the solids fraction is close to ϕ_{max}, the effective particle viscosity becomes infinite. Scrutiny of Fig 4.2(a) and Fig 4.2(b) indicates that in a dense system the contribution of the solids fraction on the stabilization of the system is mainly due to its effect on the particle stress. To further investigate the effect of these two terms, the functional forms proposed by Glasser et al. (1996; 1997) are mentioned here for comparison. In their model, the particle pressure and the particle viscosity are independent on the granular temperature, taking the form

$$P_s = C_1\phi^3 \exp(\frac{r\phi}{\phi_{max}-\phi}) \tag{4.7}$$

$$\mu_s = \frac{A\phi}{[1-(\phi/\phi_{\max})]^{1/3}} \tag{4.8}$$

where C_1, r and A are positive constants. In Equation 4.7 and 4.8, the particle pressure and the particle viscosity monotonically increase with the increase in the solids fraction and become infinite as the solids fraction is close to $\phi_{\max}$. In addition, the quantitative predictions between these two models exhibit big differences, especially for dense systems. For example, if ϕ is assumed to be equal to 0.6, the particle pressure and particle viscosity calculated from Glasser's (1997) model are 8.53 Pa and 11.39 kg/(m·s), which are much larger than the values calculated from Lun's (1984) model (see Fig 4.2). Therefore, using Glasser's (1997) model the uniformly fluidized beds are much easier to be stabilized.

To explain why Lun's model has a bounded value for the particle pressure at $\phi_{\max}$ and why the kinetic theory predicts small values of the particle pressure and particle viscosity in dense systems, a further study of the model equations and the closure functions is required. In the kinetic theory model, the main contribution of the solids faction to the particle pressure and the effective particle viscosity comes from the radial distribution function, $1/[1-(\phi/\phi_{\max})]^{1/3}$, which becomes infinite as the solids fraction is close to $\phi_{\max}$. Since the particle pressure is finite at $\phi_{\max}$, its correlation must include a term that decreases with an increase in the solids fraction and reduces to zero when the solids fraction is close to $\phi_{\max}$. To find this term, the contribution of the granular temperature is studied. Fig 4.3 shows that the granular temperature decreases with an increase in the solids fraction because a decrease in the free path between particles leads to a decrease in the fluctuational velocity as the solids fraction increases. By analyzing

Equation 4.3, it is found that with an increase in the solids fraction, the production of the pseudo-thermal energy decreases and the dissipation of the pseudo-thermal energy increases, leading to a decrease in the granular temperature. At $\phi_{\max}$ the only solution admitted in Equation 4.3 is a zero value of the granular temperature. Therefore, it is clear that the finite value of the particle pressure at $\phi_{\max}$ in the kinetic theory model is due to the automatic compensation between the contribution of the solids fraction and the contribution of the granular temperature. For the effective particle viscosity, it seems that the contribution of the granular temperature cannot completely compensate the contribution of the solids fraction. This is because the effective particle viscosity is proportional to the square root of the granular temperature while the particle pressure is proportional to the granular temperature. It is important to note that the term representing the contribution of the gas-particle slip on the pseudo-thermal energy transport, J_{vis}, is not considered in most of previous work (Ocone et al., 1993; Pita and Sundaresan, 1991; Sinclair and Jackson, 1989; Yasuna et al., 1995). Without this term, the granular temperature is always equal to zero for the uniformly steady state solution, leading to a zero value of the particle pressure and the effective particle viscosity. Therefore, under this situation the uniformly fluidized beds are always unstable and the growth rate increases monotonically with a decrease in the perturbation wavelength.

When k_x is equal to zero, the growth rate always has a negative value with the variation of k_y, indicating that the uniform state is always stable if the disturbances are only in the lateral direction (see Fig 4.4). This is qualitatively in agreement with the experimental measurements by El-Kaissy and Homsy (1976) and Didwania and Homsy (1981) that the initial instability from the uniform state is always vertical waves, and subsequently this primary wave train may lose its stability to the transverse waves. From the simulations, we found that for one-dimensional vertical perturbations the dominant instability mode has a form of traveling waves. The density layer trains move upward if ω has a positive value or move downward as ω has a negative value. However, if the perturbation is only in the lateral direction, ω is always equal to zero, indicating that the transverse density waves are stationary in gas-fluidized beds.

The stability analysis for two-dimensional disturbances is also studied in the present work. Fig 4.5(a) shows the variations of the growth rate with k_y when k_x is equal to 0, 1, 2 and 4 (m^{-1}) respectively. With two-dimensional disturbances, there are two branches in the growth rate profile. The growth rate starts from a maximum with k_y equal to zero, and monotonically decreases with an increase in k_y. When the value of k_y passes through a certain number, the growth rate transits to the second branch. It is interesting to note that the second branch is identical with the branch of k_x equal to zero, indicating that the transverse waves dominate the stability behavior under this situation. This transition is also revealed in the phase velocity profiles. We have known that the vertical instability has a form of traveling waves and the transverse waves are stationary in the bed. In Fig 4.5(b), the phase velocity has a nonzero value corresponding to the first branch (in this specific case, the traveling waves move downward in the bed), indicating that the vertical disturbance plays a key role in the system instability. With a further increase in k_y, the dominant mode in the growth rate profile transits to the second branch, and a zero value is obtained in the corresponding phase velocity profiles, indicating that the lateral disturbance becomes important. From the transition of the dominant modes observed in Fig 4.5, we can see that the stability behavior is much more sensitive to the variation of k_x

than that of k_y. Since both the wavelength and the growth distance are inverse to the wave number, the vertical disturbances should grow much faster than the lateral disturbances, and transverse structures should not be observed in a short and narrow bed.

The eigenfunctions associated with the solids fraction are shown in Fig 4.6. In the contour map, the alternating bands of dense and dilute regions are inclined at an angle to the axis. The value of the angle is determined by the ratio of k_x and k_y. It can be imagined that if k_y is zero, a density wave train with a slug pattern can be observed to move through the bed.

Finally, a parametric analysis is investigated on the system. From Fig 4.1 and Fig 4.4 we have known that the solids fraction has a positive effect on the stabilization of the gas-particle flow system. The next parameter considered is the coefficient of particle-particle restitution (e_p). Previous works have reported that the inelasticity alone can lead a granular system to lose its stability to the emergence of relative dense clusters (Goldhirsch, 1999; Goldhirsch and Zanetti, 1993; Tan and Goldhirsch, 1997). For example, a simple homogeneous and unforced granular system may be "cooling" automatically due to the system inelasticity. Assuming the particle concentration has a small increase in a region due to a flucturation, the granular temperature will decrease faster in this region than in the neighboring regions because the frequency of inelastic collisions is larger in this region than in the neighboring, less dense region. This leads the particle pressure to form a gradient from the neighborhood to this dense region, causing a migration of more particles into this region, thus further increasing its density and decreasing its pressure. Therefore, a dense cluster is formed by the inelasticity and a particle concentration fluctuation. Since the inelasticity plays an important role in the instability of granular systems, an investigation of its effect on the instability of gas-particle mixed flows is required. The contribution of the coefficient of particle-particle restitution is shown in Fig 4.7. To get better comparison, the range of k_x is much smaller than k_y since the instability of unbounded gas-particle flows is much more sensitive to the variation of k_x than k_y. In the figures, the line labeled with zero is a line of demarcation. On its right side the maximum value of ξ for a pair of k_x and k_y is negative, indicating the system is stable, while the area on its left side represents a positive value of ξ, indicating the system is unstable. With increasing the value of e_p, the stable area increases, indicating that the system is easier to be stabilized. This result qualitatively agrees with previous works (Hirayama and Tsuzuki, 1995). It is important to note that using this kinetic theory model, the unbounded uniform gas-particle fluidized beds is always unstable to vertical wave disturbances, even if the collisions between particles are assumed elastic.

4.2 Linear Stability Analysis for Bounded Gas-Particle Fluidized Beds

Before analyzing the instability in bounded gas-particle fluidized beds, we need to recall their steady state solutions as the starting point. The base physical parameters used in the simulations in this chapter are listed in Table 4.1.

4.2.1 Steady State Solutions

With the adding walls, the base state of the gas-particle flows is not uniform. Instead, the segregation of particles can be observed from the simulations. The governing equations as well as boundary conditions for steady, fully developed flows have been presented in Chapter 3.

This work focuses on gaining a better understanding of the effect of walls on unstable flow behavior in gas-particle systems having spatially non-uniform steady states. We start by investigating the effect of the coefficient of particle-wall restitution (e_w) since it is known that inelasticity has a significant effect on system instability (Wang and Tong, 2001). Wang et al. (1996; 1997) reported that the presence of walls in granular flow within a channel could serve as either a *source* or a *sink* of fluctuation energy depending on the value of e_w and e_p.

Base state profiles of solids volume fraction (ϕ), (dimensionless) granular temperature (T^*), and (dimensionless) gas and particle velocities (u^* and v^*) for the source and sink wall conditions are shown in Figs. 4.8 and 4.9, respectively. To better investigate the effect of wall properties, the value of e_p is held constant at 0.96, and the value of e_w is varied from 0.6 for the sink wall condition to 0.98 for the source wall condition. When the wall is a *source* of fluctuation energy, we see in Fig. 4.8(a) that ϕ monotonically decreases when moving away from the center towards the walls and that this corresponds to an increase in granular temperature, T^* (see Fig. 4.8(b)). If the average solids fraction is high enough, a plug flow can be observed where $\phi=\phi_{max}$ at the center. Variations in u^* and v^*, for the source wall condition are shown in Fig 4.8(c) and 4.8(d) respectively. In Fig. 4.8(d), v^* reaches a maximum value in the center, and monotonically decreases towards the wall. In Fig. 4.8(c), u^* is shown to initially increase as one moves away from center, and then passes through a maximum value corresponding to high values in the granular temperature profile. As the wall is approached, u^* is shown to rapidly decrease.

In Fig. 4.9(a) we see that the solids volume fraction profile for the wall *sink* condition is very similar to the source wall case in Fig. 4.8(a), except in the region very near the wall where a slight increase in ϕ is observed. This can be explained by the fact that the granular temperature decreases close to the walls in this case. T^* is shown to increase away from the center, passes through a maximum and finally decreases rapidly near the walls. Profiles of v^* and u^* are very similar to the source wall case except that the maximum value of u^* is more pronounced under the source condition (Fig. 4.8(c)) than the sink condition (Fig. 4.9(c)) due to higher values of T^* near the walls (see Fig. 4.8(b) versus Fig. 4.9(b)).

We have shown that profiles of solids volume fraction are very similar for both sink and source wall cases in Figs. 4.8(a) and 4.9(a) with the exception of the region adjacent to the walls. Hence, e_w plays only a minor role within the range of 0.6 to 0.98 as shown. However, the similarity in profile is strongly dependent on the value of e_p as shown in Fig. 4.10(a) and (b) where profiles for ϕ and T^*, respectively, are plotted for the e_w values of 0.6 and 0.99 at a value of e_p of 0.99. We see in this figure that for e_p very close to 1, profiles for ϕ and T^* differ qualitatively and quantitatively, as one approaches the wall. This is particularly evident for the sink wall condition in Fig. 4.10(a) where the increase in ϕ near the walls becomes so pronounced that particles equally segregate both to the center and at the walls. This is unlike the source wall case where particle concentration monotonically decreases from the center to the walls. Hence, particle concentration at the center is much higher under the source wall condition than the sink wall condition. We find that if both e_p and e_w are equal to 1, a solution with uniform solids fraction can be obtained from simulations.

4.2.2 The Linear Stability Approach

To investigate the stability of the above steady state solutions, a technique of linear stability analysis is used on the governing equations as well as the boundary conditions. The detailed analysis procedure is similar to the unbounded case except that the trial solution for the perturbations has a different form. The solutions for the two dimensional perturbations are considered to take the following form

$$\underline{u_1} = \underline{\hat{u}_1(y)}\exp(st)\exp(ik_x \cdot x)\ , \qquad \underline{v_1} = \underline{\hat{v}_1(y)}\exp(st)\exp(ik_x \cdot x)$$

$$\phi_1 = \hat{\phi}_1(y)\exp(st)\exp(ik_x \cdot x), \qquad P_1 = \hat{P}_1(y)\exp(st)\exp(ik_x \cdot x)$$

$$T_1 = \hat{T}_1(y)\exp(st)\exp(ik_x \cdot x) \tag{4.9}$$

where $\underline{\hat{u}_1(y)}$, $\underline{\hat{v}_1(y)}$, $\hat{\phi}_1(y)$, $\hat{P}_1(y)$, and $\hat{T}_1(y)$ represent the amplitudes of the perturbations for the gas velocity, particle velocity, solids volume fraction, gas pressure and granular temperature respectively. They are functions of the lateral space instead of constants in unbounded cases. k_x, a real component, is the wave number of the plane wave disturbance in the vertical direction. Substituting the proposed form of disturbance into the linearized equations with a given value of k_x, a set of ordinary differential equations of perturbation amplitudes could be obtained (see Appendix 3, where the solutions of the base state are described by adding a superscript zero).

This eigenvalue problem is solved by discretizing the equations using a second-order finite difference approximation. A polynomial function is used to approximate the steady state solutions, based on the known numerical simulation results from the previous section. Typically the interval (Zhou et al., 2002) for the lateral direction is divided into n equal sub-intervals, and $\hat{u}_1(\text{y})$, $\hat{v}_1(\text{y})$ and $\hat{T}_1(\text{y})$ are evaluated at the $n+1$ grid points (including the two boundary points). $\hat{\phi}_1(\text{y})$ and $\hat{P}_1(\text{y})$, on the other hand, are evaluated at the mid points of the sub-intervals because it is difficult to find appropriate boundary conditions for these two variables. This "staggered" form allows us to reduce the number of unknowns by two. Therefore, the numbers of equations and variables become the same. Then the ordinary differential equations are discretized, leading to a matrix eigenvalue problem, which take the form

$$AX = sBX \tag{4.10}$$

Where A and B are square matrix with $7n$-2 by $7n$-2 dimensions. Since the boundary conditions do not contain the eigenvalue s and the coefficients of s in the continuity equations for the gas phase can be eliminated by those in the continuity equations for the solids phase, B is a singular matrix. This singularity can be removed by row and column operations. Finally, this eigenvalue problem is solved using the Matlab6.5 software.

The accuracy of the numerical scheme is ascertained by increasing the number of collocation points until the leading eigenvalue converges to a value that is independent with the variation of the number of collocation points. It is found that the larger the value of k_x, the larger is the number of collocation points required to achieve convergence. To ensure the accuracy of the numerical approach we used, our simulation code is checked by favorable comparison with published results by Wang C.H (1995) and Alam and Nott (1998). For instability analysis of bounded gas-particle fluidized beds, our results are exhibited based on dimensionless analysis (in this section, k_x represents a dimensionless vertical wave number).

4.2.3 Instability in Gas-Particle Fluidized Beds with Sink Walls

When e_w is small, the walls act as an energy sink for fluctuation energy. Under these conditions, the stability structure map in the $\bar{\phi}$-e_p plane (average solids volume fraction – coefficient of restitution) is shown in Fig 4.11. In this work, the eigenvalue (dimensionless) s having maximum positive growth rate, ξ, at a vertical wavenumber (dimensionless) k_x is referred to as the *leading* eigenvalue, and s having maximum ξ for all k_x is referred as the *dominant* eigenvalue. Within the physical parameter space, the stability analysis shows the dominant mode in the $\bar{\phi}$-e_p plane has three types of structures: stable patterns; symmetric unstable patterns; and anti-symmetric unstable patterns.

The upper curve in Fig 4.11 represents the boundary between the stable and unstable regions. We can see in this figure that there is a monotonic increase in the region of stability at increased values of e_p as well as increased average solids volume fraction, $\bar{\phi}$. Recall that the base state transitions from a "Core-Annular" to a "Core-Segregation" structure when the value of e_p decreases from 1. From a physical standpoint, this contour map suggests gas-fluidized beds are more easily stabilized under densely packed conditions (i.e., for $\bar{\phi} \rightarrow \phi_{max}$), or when particle-particle collisions are highly elastic (i.e., $e_p \rightarrow 1$), such that the bed acquires a "Core-Annular" structure. As shown in Fig 4.11, as $\bar{\phi}$ decreases in the unstable region, the dominant instability mode transitions from a symmetric to anti-symmetric structure.

In order to better understand the sensitivity of mode transition to particle packing, three ($\bar{\phi}$,e_p) points were selected in the unstable region of Fig 4.11: (0.1,0.96); (0.15,0.96); (0.2,0.96). The profiles of (dimensionless) growth rates (of the leading eigenvalues) and their corresponding (dimensionless) phase velocities are plotted in Fig 4.12. As illustrated in Fig. 4.12 (a), growth rate, ξ, has two branches, which cross at a k_x value located at the point labeled **I.** At this point, there is a corresponding step change in the phase velocity profile as shown in Fig. 4.12 (b). The first branch labeled **S** in Fig. 4.12 (a) represents all symmetric mode structures and the second branch labeled **A** represents all anti-symmetric structures. Fig. 4.12 (a) shows that the maximum growth rates of both symmetric and anti-symmetric branches increase with a decrease in the solids fraction. When $\bar{\phi}$ is equal to 0.2, the maximum growth rate of the dominant eigenvalue is 0.121 for the symmetric structure at k_x=0.5, and the maximum growth rate of the anti-symmetric branch is –0.017 at k_x=2.5. Since the maximum growth rate of the second branch is negative, the system is stable to the anti-symmetric disturbances. When $\bar{\phi}$ decreases to 0.15, the maximum growth rates of the symmetric and anti-symmetric branches increase to 0.180 and 0.181 at k_x=0.65 and 2.5 respectively, indicating that the growth of the anti-symmetric perturbations is much faster than the symmetric perturbations with a decrease in $\bar{\phi}$. In this case, the system is equally unstable to the symmetric and anti-symmetric disturbances. With a decrease in $\bar{\phi}$ to 0.1, the dominant mode transitions to the anti-symmetric pattern, where the maximum growth rate of the dominant eigenvalue is 0.386 at k_x=2.5. The profiles of the phase velocities corresponding to the leading eigenvalues are shown in Fig 4.12 (b). The phase velocity changes very slowly and smoothly with respect to k_x provided a mode transition is not observed in the corresponding eigenvalue profile. Note that as $\bar{\phi}$ decreases from 0.2 to 0.1, the phase velocity corresponding to the maximum growth rate in the symmetric branch increases from 10.72 to 11.60, while phase velocity (corresponding to maximum growth rate) for the anti-symmetric branch increases from 9.57 to 9.92. Moreover, the magnitude of the phase velocities of the disturbances are slightly greater than the solid phase velocities, v^*, shown in velocity profiles (Fig. 4.9(d)) at k_x values corresponding to maximum growth rates, but significantly less than gas phase velocities, u^*, (shown in Fig. 4.9(c) profiles). These results suggest that the propagation of the disturbance waves is primarily dominated by the solids phase. The similarity in magnitude of v^* and the disturbance phase velocity was also observed in granular flow. Wang et al. (1997) reported that eigenmodes in granular flow within a channel are traveling waves, which move with a speed comparable to, but slightly greater than, that of the granular material itself.

The eigenfunctions corresponding to the dominant eigenvalues are plotted in Figs. 4.13 and 4.14 for average solids volume fractions, $\bar{\phi}$=0.2 and 0.1 respectively. Eigenfunctions associated with solids volume fractions are plotted in Figs. 4.13(a) and 4.14(a), and eigenfunctions associated with solids phase velocity are plotted in Figs. 4.13(b) and 4.14(b). In all figures, the eigenfunctions are plotted as spatial variations over one periodic cell of height $2\pi/k_x$. In both cases for $\bar{\phi}$, e_p and e_w are set at 0.96 and 0.6 respectively. For the denser of the two cases examined, ($\bar{\phi}$=0.2), Fig. 4.13(a) shows that the dominant mode has the form of symmetrical structures where particle density patterns are symmetric about the centerline of the channel (y*=0). This means that traveling waves propagate upwards through the column with alternating layers of dense and dilute regions. The eigenfunctions associated with the solid phase velocity are shown in Fig 4.13(b) as a vector plot. It can be seen that over one periodic cell, the velocity profile vertically breaks down into two distinct regions corresponding to the distribution of the alternating density layers in Fig. 4.13(a). In particular, particle velocity has an upward fluctuation in dense layers and a downward fluctuation in dilute layers. Combining Fig 4.13(a) and (b), we can simply describe the motion of particles in gas-fluidized beds with a symmetrical perturbation in the following way: when the lower surface of the dense region becomes unstable, particles (within the solids rich region) "rain" down to fill the dilute region below. Once they pass through the dilute region, they enter the upper surface of another solids rich region. Hence, for any particle dense region, particles are lost from its lower surface and simultaneously collected on its upper surface. It is this mechanism that leads to the propagation of the density wave. A similar mechanism has been proposed by Glasser et al. (1998) to explain the formation of bubbles and clusters in fluidized beds.

In the more dilute case ($\bar{\phi}$=0.1), the dominant mode shown in Fig. 4.14(a) has an anti-symmetric structure with alternating regions of high and low densities positioned on either side of the centerline. Although eigenfunctions associated with the solids phase velocity vectors in the vertical direction (Fig. 4.14(b)) appear similar to the symmetric case (where particles accelerate upwards in dense regions and downwards in dilute regions), lateral fluctuation becomes important in the anti-symmetric structures because the particle concentration has a significant gradient across the centerline. This is not the case for the symmetric structure as previously noted. As a result of the vertical and lateral components, recirculation loops are observed close to the centerline where particles migrate from a dense (dilute) to a dilute (dense) region.

It is clear that there exist two types of disturbance structures in the gas-fluidized system, and that the leading instability mode can transition from one to the other at some value of k_x. However, whether this transition is smooth is still not understood. To study the transition characteristics, the contour maps of leading mode eigenfunctions associated with solids fraction are shown at several k_x points for a specific case ($\bar{\phi}$=0.15, e_p=0.96). With the variation of k_x, the eigenfunctions corresponding to the maximum growth rate for two instability branches are shown in Fig 4.15(a) and (d). It is clear that the first branch has symmetric patterns, while the second one represents anti-symmetric patterns. Fig 4.15(b) and (c) show the eigenfunction plots just before the transition and just after that, respectively. It can be seen that there is a significant change in the natures of the solids fraction structures. The transition observed is not smooth, indicating that these two disturbance patterns have no internal relations and they independently develop with the variation of k_x.

Another parameter we have considered in this work is the coefficient of particle-particle restitution, e_p. In Fig 4.11, it was previously shown that all observed instabilities are driven by the inelastic nature of particle-particle collisions. In particular, the perfectly elastic system (e_p =1) is always stable to small perturbations, and systems become less stable to small disturbances as e_p decreases from a value of one. It is interesting to note that for a range of $\bar{\phi}$ values from 0.10 to 0.15, the unstable modes have two transition points with a decrease in e_p; the first represents a transition from a symmetric to anti-symmetric structure; and the second represents a reverse transition back to a symmetric pattern (see Fig. 4.11). To further investigate the origin of these two transition points, growth rates and phase velocities of the disturbance are plotted as a function of wavenumber, k_x for solids volume fraction $\bar{\phi}$=0.15 at various values of e_p in Fig. 4.16. In Fig. 4.16(a), as e_p decreases from 0.98 to 0.965, the instabilities in both branches are significantly enhanced, and the leading eigenvalue in the anti-symmetric branch becomes larger than that in the symmetric branch. Hence, there exists a transition from a symmetric to anti-symmetric pattern under these conditions. However, as e_p decreases beyond this point, i.e., from 0.965 to 0.95, the potential exists for the dominant instability mode to transition back to the symmetric pattern since, within this range of e_p values, the growth of the leading eigenvalue in the symmetric branch becomes larger than that of the anti-symmetric branch. A comparison of Figs. 4.12 and 4.16 shows that with decreasing $\bar{\phi}$, the k_x point corresponding to the transition of the two leading eigenvalue branches shifts to the left allowing the anti-symmetric branch to become dominant within a larger range of k_x values. Conversely, as e_p decreases, the range of k_x values at which the anti-symmetric branch dominates is restricted.

4.2.4 Instability in Gas-Particle Fluidized Beds with Source Walls

We now consider the stability analysis for the case of source walls, when e_w increases to 0.98. To compare with the stability results from the analysis of sink walls (Fig 4.12-4.14), Two situations, ($\bar{\phi}$=0.1, e_p=0.96, e_w=0.98) and ($\bar{\phi}$=0.2, e_p=0.96, e_w=0.98), are selected for investigation. Fig 4.17 shows that for source walls the growth rate profiles still have two branches. The first one represents symmetric patterns and the second one represents anti-symmetric patterns. It is important to note that the instability patterns under source walls conditions are very similar to those under sink wall conditions. Only the transition point between two eigenvalue branches and the associated jump in phase velocity profiles have a slight difference. This insensitivity of stability behavior to the nature of boundary walls is also shown in the eigenfunction contour plots. Fig 4.18 shows the solids fraction eigenfunctions corresponding to the dominant eigenvalues at ($\bar{\phi}$=0.1, e_p=0.96, e_w=0.98) and ($\bar{\phi}$=0.2, e_p=0.96, e_w=0.98), respectively. Scrutiny of Fig 4.13, 4.14 and 4.18 reveals that the contour maps of eigenfunctions associated with solids fractions are almost indistinguishable between the sink wall and source wall cases.

We believe this insensitivity of stability features to the nature of boundary conditions is due to the similarity of solids fraction profiles in the base state (see Fig 4.8(a) and Fig 4.9(a)). When e_p is smaller than 0.99, particles are always strongly segregated at the center and the contribution of e_w on the particle distribution is small. When e_p is larger than 0.99, the discrepancy of solids fraction profiles between sink wall conditions and source wall conditions become pronounced (see Fig 4.10(a)). However, from Fig 4.11 it can be seen that when e_p is close to 1, the dominant eigenvalue becomes negative and the gas fluidized bed system is stable to small perturbations. Since the disturbances will damp out in this case, this situation is not of interest.

4.2.5 Time-Dependent Evolution of Linear Instabilities

Finally, the steady state solution is augmented with a small perturbation to examine its spatio-temporal effect according to $\phi_t = \hat{\phi}_1 \exp(st^*)\exp(k_x x^*) + \phi_0$, where ϕ_0 is the base state solids fraction and ϕ_t is the time-dependent solids fraction. Fig 4.19 shows the results for a system dominated by the anti-symmetric perturbations. From time equal to zero, a streamer-like solution forms in the center due to the instability patterns (see the eigenfunction plot in Fig 4.19), and moves upwards in a "snake-like" fashion, oscillating to a greater extent about the center with time. When the dominant eigenfunctions are symmetric, the evolution of the instability behavior is markedly different (see Fig 4.20). Since particles are initially segregated at the center in the base state condition and dominant instabilities are also observed at the center (see the eigenfunction plot in Fig 4.20), a cluster-like structure can be captured immediately. In addition, with increasing time the low density regions (located near the walls in the base state) migrate to the center and coalesce to form a compact region of low density. At this point, alternating regions of low and high density can be seen moving throughout the bed. Finally, the pattern develops in a bubble-like structure. More time is required to capture the formation and propagation of bubble-like structures due to the solids distribution of the base state.

From Fig 4.19 and 4.20, it can be seen that there is little to no change in density structure in regions near the wall, and that most of the changes are observed at the center. Therefore, it is clear that the boundaries serve to stabilize the system, and that instability structures develop in the center region.

4.3 Physical Mechanism of Instabilities

The physical mechanism of instability formation and propagation discussed in the previous section is examined by analyzing the governing equations term-by-term.

4.3.1 Effect of the Inertial Terms

It is well known that the instability of fluidized beds is of an inertial nature (Hirayama and Tsuzuki, 1995; Jackson, 1963). Hirayama and Takaki (1995) reported that for an unbounded liquid-fluidized bed, the fluid inertial term is responsible for the instability. In gas-particle beds however, fluid phase inertia is very small (and often neglected) Hence, its inertial contribution to instability formation should be much smaller than in liquid-fluidized beds, while the contribution of solids phase inertia should be more pronounced.

To study the effect of the inertial terms on the mechanism of the instability, the inertial terms in the momentum balance equations are multiplied by an empirical coefficient α, which can vary from 1 to 0. The variation of the leading eigenvalue and its corresponding phase velocity with k_x is then calculated. The eigenvalue profiles of dominant growth rate and particle phase velocity are subsequently obtained for specific values of α. Since the inertial terms go to zero in the steady state equations for fully developed flows, the coefficient α does not change the base state solution.

Fig 4.21 shows the dominant eigenvalue and its phase velocity plotted as a function of α for a system dominated by the anti-symmetric instability mode ($\bar{\phi}$=0.1, e_p=0.96, e_w=0.6). Case I considers the effect of the gas inertial term, which is multiplied by α in the gas phase momentum equations; Case II considers the effect of the solid phase inertial term, which is multiplied by α in the solids phase momentum equations. The effect of both the gas and solid phase inertial terms is considered in Case III.

Although the growth rate is shown to decrease in all three cases examined, as the value of α decreases from 1, there are still some significant differences worth noting. In Case I, a plateau is observed when α decreases from 1 to 0.65, indicating that (within this region) the effect of the gas inertial term on the instability nature is quite small. In Case II, as α decreases from 1, growth rate first decreases, passes through a plateau ($0.7 < \alpha < 0.4$), and then decreases rapidly to a negative value. The plateaus observed in Cases I and II is not observed in Case III, where growth rate monotonically decreases with a variation of α from 1 to 0. The transition of the dominant instability pattern with the variation of α can be found in Fig 4.21(b), where the symmetric pattern corresponds to the high velocity branch, and the anti-symmetric pattern corresponds to the low velocity branch. In Case I, the instability pattern is always dominated by the anti-symmetric structure. In Case II, as α is decreased from one, the dominant instability pattern initially has an anti-symmetric form, and then the dominant inertial force transitions from the solids phase to the gas phase. Close examination of Figs. 4.21(a) and (b) reveals that, in this case, the gas inertial term eventually leads to the development of the symmetric instability pattern and that the solid phase inertial term is important for the development of the anti-symmetric instabilities. Therefore, when anti-symmetric patterns dominate system instability formation, the contribution of the solid phase inertial term is more pronounced.

When the system instability is dominated by symmetric patterns ($\bar{\phi}$=0.2, e_p=0.96, e_w=0.6), the variation of the dominant growth rate and its phase velocity with α is shown in Fig 4.22, where the description of Case I – Case III is same with that in Fig 4.21. The high phase velocity branch in Fig 4.22(b) represents the instability dominated by symmetric patterns, while the low phase velocity branch denotes the instability dominated by anti-symmetric patterns. In Case I, with decreasing α from 1 to 0.55 the growth rate decreases and symmetric patterns dominate the system instability. As α continuously decreases, the dominant instability transits to anti-symmetric patterns, and the growth rate starts to increase. Then it passes through a maximum and decreases quickly as α is close to 0. This non-monotonic behavior has not been observed in previous works. In Case II, a plateau can be observed when α decreases from 1 to 0.55, indicating a small contribution of the solids inertial term in this range. With further decreasing α, the grow rate starts to decrease, and a negative growth rate can be obtained when α is close to zero. From Fig 4.22(b), it can be seen that the dominant instability always has symmetric patterns in Case II. As both inertia terms are considered in Case III, the gas inertial force dominates the instability natures when α decreases from 1 to 0.55 because in this range the growth rate curves in Case I and Case III are almost overlapped. With continuously decreasing α, the effect of the solid inertial terms becomes important and the dominant instability has a transition from symmetric patterns to anti-symmetric patterns.

Although the growth rate profiles look quite different between Fig 4.21 and Fig 4.22, several common features still can be observed. First, the development of the symmetric patterns is mainly due to the contribution of the gas inertial terms, while the development of the anti-symmetric patterns is mainly due to the contribution of the solid phase inertial terms. Second, the origin of the instability is the inertia of the solids. Without this term, gas-fluidized beds are always stable to small perturbations. Similar results were also obtained from the work of Wang and Tong (2001). They found in a granular channel flow driven by gravity, that all modes of instability are of a solids inertial origin, and that the dominant eigenvalue monotonically decreases with a decrease in the solid phase inertial term. The contribution of the gas inertial terms on the origin of the instability is less important. The fluidized bed system may still lose stability to small disturbances when this term is absent. Third, variations of the gas and solid phase inertial terms have very little effect on the value of the phase velocity. Scrutiny of Fig 4.21(b) and 4.22(b) reveals that phase velocity changes very slowly with a variation in α if there is no transition in the dominant instability mode.

From Fig 4.21 and 4.22, it is clear that the system inertial terms have a significant effect on the instability patterns. Therefore, a further investigation of the inertial force profiles across the bed is required.

$$gas\_x = DE(1-\phi^0)(s\hat{u}_x + ik_x u_x^0 \hat{u}_x + \hat{u}_y \frac{du_x^0}{dy*}) \tag{4.11}$$

$$gas\_y = DE(1-\phi^0)(s\hat{u}_y + ik_x u_x^0 \hat{u}_y) \tag{4.12}$$

$$solids\_x = D\phi^0(s\hat{v}_x + ik_x v_x^0 \hat{v}_x + \hat{v}_y \frac{dv_x^0}{dy*}) \tag{4.13}$$

$$solids\_y = D\phi^0(s\hat{v}_y + ik_x v_x^0 \hat{v}_y) \tag{4.14}$$

where Equation (4.11) and (4.12) represent the dimensionless gas phase inertial forces in the vertical direction and lateral direction respectively, and Equation (4.13) and (4.14) represent the dimensionless solids phase inertial forces in the vertical direction and lateral direction respectively. When the instability is dominated by the symmetric patterns (see Fig 4.23(a)), the inertial forces in the vertical direction are symmetric between the centerline, while the profiles of the inertial forces in the lateral direction are anti-symmetric. In addition, the inertial forces in the vertical direction (x) are much larger than the lateral direction (y). Thus the velocity fluctuation is mainly in the x-direction (see Fig 4.13(b)). In x-direction, although the absolute value of the solids phase inertial force is larger than the gas phase, the difference is not significant and their absolute values have the same order across the bed except in the region near the center. This is mainly attributed to the gas velocity and the solids fraction profiles in the base state because a big jump can be observed in the gas velocity profile close to the walls where the solids fraction has a low level. Therefore, the gas phase inertial force has a pronounced effect on the symmetric instabilities. When the instability is dominated by the anti-symmetric patterns (see Fig 4.23(b)), the inertial forces in the vertical direction are anti-symmetric between the centerline, while the profiles of the inertial forces in the lateral direction are symmetric. The inertial forces of the gas phase are much smaller than the solids phase. Thus, the anti-symmetric instability is dominated by the solids phase inertial forces. In addition, it is found that the values of the solids inertial forces have the same order in x-direction and y-direction. Therefore, the lateral fluctuation in velocity profiles becomes important and recirculation loops can be observed in the center (see Fig 4.14(b)).

4.3.2 Effect of the Dissipation Terms

The effect of e_p on the instability nature has been mentioned in the previous sections. Since the constitutive relations used in this model is based on slightly inelastic particles, the contribution of e_p mainly acts on the dissipation term of particle-particle collisions (J_{coll}) in the pseudo-thermal energy balance equation. To understand the effect of this term on the mechanism of instability, an empirical coefficient α is added onto this term and its value is varied from 1 to 0.

It is important to note that a decrease in the dissipation term has a significant effect on the profiles of the base state. Fig 4.24 shows that with a decrease in α, the segregation of particles has a transition from the center to the walls. When this term is eliminated, a "Core-Annular" flow is observed. Therefore, it is clear that the sensitivity of steady state flow structures to e_p is due to the fast increase in the dissipation term as the system transits from an elastic case to a slightly inelastic case. If the contribution of the dissipation term is reduced, this sensitivity can be suppressed. This topic has been discussed in many papers (Eduardo et al., 1995; Hrenya and Sinclair, 1997; Louge et al., 1991; Ocone et al., 1993; Pita and Sundaresan, 1991; Sinclair and Jackson, 1989).

Under the situation of $\overline{\phi}$=0.2, e_p=0.96 and e_w=0.6, the variation of the dominant growth rate and its phase velocity with α is shown in Fig 4.25. It is found that the maximum growth rate in both symmetric and anti-symmetric branches decreases with the decrease in α, and the decrease is much faster in the anti-symmetric branch. Since symmetric patterns dominate the system instability when α is equal to 1, the dominant instability always has symmetric patterns as α decreases from 1 to 0. It is important to note that the dissipation term has a great effect on the system instability. When α reduces to 0.4, the dominant growth rate becomes negative and the gas-fluidized bed is said to be stable. This effect is also shown in the low solids fraction situation ($\overline{\phi}$=0.1, e_p=0.96 and e_w=0.6), where the system becomes stable when α is below 0.25 (see Fig 4.26). Fig 4.26(b) shows that the dominant instability pattern has a transition from anti-symmetric structures to symmetric structures as α decreases. This is in agreement with the previous finding that the decrease in the dominant growth rate is much faster in the anti-symmetric branch than in the symmetric branch with decreasing α. By examining Fig 4.21, 4.22,

4.25 and 4.26, we can conclude that the origin of instability must include both solids inertial and granular energy dissipation terms, otherwise the system is stable to small perturbations.

4.4 Effects of Flow Hydrodynamics on Chemical Reactions

In gas-particle chemical reactors, high yield and selectivity are critical for the competitiveness of a process and for minimizing environmental impacts. The different flow regimes and instabilities lead to significantly different mixing characteristics, and therefore influence selectivity and yield of chemical reactions. Low yield and selectivity result in the formation of large quantities of byproducts requiring separation, disposal or rework.

4.4.1 Reaction Model and Numerical Method

To study the effects of flow hydrodynamics on chemical reactions, a simple chain reaction is assumed to occur in a gas-particle fluidized bed.

$$A \xrightarrow{Catalyst} B \qquad\qquad B \longrightarrow S(byproduct) \tag{4.15}$$

where A is the raw material, B is the target product and S is the byproduct. In this model, we assume that A, B and S are in the gas state in the reactor and the catalyst is a solid (particles). Therefore, the chemical reactions do not affect the flow profiles, and the hydrodynamic results we obtain in previous chapters can be still used here. The equations representing the balance of concentrations, C_A, C_B, and C_s, take the form (Bird et al., 1960):

$$\frac{\partial C_A}{\partial t} + \underline{u} \cdot \nabla C_A = D_A (\frac{\partial^2 C_A}{\partial x^2} + \frac{\partial^2 C_A}{\partial y^2}) + R_A \quad (4.16)$$

$$\frac{\partial C_B}{\partial t} + \underline{u} \cdot \nabla C_B = D_B (\frac{\partial^2 C_B}{\partial x^2} + \frac{\partial^2 C_B}{\partial y^2}) + R_B \quad (4.17)$$

$$\frac{\partial C_S}{\partial t} + \underline{u} \cdot \nabla C_S = D_S (\frac{\partial^2 C_S}{\partial x^2} + \frac{\partial^2 C_s}{\partial y^2}) + R_S \quad (4.18)$$

where $\underline{u}$ is the superficial velocity of the gas phase, D_A, D_B, and D_s are ordinary diffusion coefficient of species A, B and S, and the reaction rate of each species can be written as

$$R_A = -k_A C_A C_{catalyst} \quad (4.19)$$

$$R_B = k_A C_A C_{catalyst} - k_B C_B \quad (4.20)$$

$$R_S = k_B C_B \quad (4.21)$$

where k_A and k_B, are constants representing the reaction rate coefficients.

In Equation 4.16 to 4.18, the gas velocity and solids fraction are obtained from the simulations in the previous chapters. Since the concentration of the catalyst is only dependent on the solids fraction, the unknowns in the concentration equations are C_A, C_B, and C_s. To solve this problem, two boundary conditions and one initial condition are required for each species. The boundary conditions at the walls are obtained by calculating the concentration balance in a thin layer close to the walls.

$$\frac{\partial C_A}{\partial t} = -u_x \cdot \frac{\partial C_A}{\partial x} - u_y \frac{C_A}{\delta} + D_A \frac{\partial^2 C_A}{\partial x^2} + D_A \frac{\partial C_A}{\partial y} \frac{1}{\delta} + R_A \quad (4.22)$$

$$\frac{\partial C_B}{\partial t} = -u_x \cdot \frac{\partial C_B}{\partial x} - u_y \frac{C_B}{\delta} + D_B \frac{\partial^2 C_B}{\partial x^2} + D_B \frac{\partial C_B}{\partial y} \frac{1}{\delta} + R_B \quad (4.23)$$

$$\frac{\partial C_S}{\partial t} = -u_x \cdot \frac{\partial C_S}{\partial x} - u_y \frac{C_S}{\delta} + D_S \frac{\partial^2 C_S}{\partial x^2} + D_S \frac{\partial C_S}{\partial y} \frac{1}{\delta} + R_S \quad (4.24)$$

where δ is a parameter related with the thickness of a layer adjacent to the wall in which the concentration balance is calculated. The initial conditions come from the steady state simulations.

For the steady state, the governing equations can be rewritten as

$$\frac{\partial C_A}{\partial x} = P_A \qquad \frac{\partial C_B}{\partial x} = P_B \qquad \frac{\partial C_S}{\partial x} = P_s \tag{4.25}$$

$$\frac{\partial P_A}{\partial x} = (u_x \cdot P_A + R_A) / D_A - \frac{\partial^2 C_A}{\partial y^2} \tag{4.26}$$

$$\frac{\partial P_B}{\partial x} = (u_x \cdot P_B + R_B) / D_B - \frac{\partial^2 C_B}{\partial y^2} \tag{4.27}$$

$$\frac{\partial P_S}{\partial x} = (u_x \cdot P_S + R_S) / D_S - \frac{\partial^2 C_S}{\partial y^2} \tag{4.28}$$

and the boundary conditions take the form:

$$C_A = C_{A0}(1-\phi) \qquad C_B = C_S = 0 \qquad \text{at } x=0 \tag{4.29}$$

$$P_A = R_A / u_x \qquad P_B = -R_A / u_x \qquad P_s = 0 \qquad \text{at } x=0 \tag{4.30}$$

$$\frac{\partial P_A}{\partial x} = (\underline{u_x} \cdot P_A \pm R_A) / D_A - \frac{\partial C_A}{\partial y}\frac{1}{\delta} \qquad \text{at } y = \pm 1 \tag{4.31}$$

$$\frac{\partial P_B}{\partial x} = (\underline{u_x} \cdot P_B \pm R_B) / D_B - \frac{\partial C_B}{\partial y}\frac{1}{\delta} \qquad \text{at } y = \pm 1 \tag{4.32}$$

$$\frac{\partial P_S}{\partial x} = (\underline{u_x} \cdot P_S \pm R_S) / D_S - \frac{\partial C_S}{\partial y}\frac{1}{\delta} \qquad \text{at } y = \pm 1 \tag{4.33}$$

where C_{A0} is the concentration of A if the bed is occupied only by this material.

The above model (Equations 4.25 to 4.33) is solved by discretizing the equations in the lateral direction and integrating them in the vertical direction, using the LIMEX approach. LIMEX is a solver based on the semi-implicit Ruler discretization (Deuflhard et al., 1987; Ehrig et al., 1999; Khinast, 2001). A major advantage of this solver is that it is less sensitive to inconsistencies in the initial conditions. For the steady state problem, the boundary conditions at x=0 can be treated as initial conditions, and the calculation in the x step can be treated as the time step required in the LIMEX solver.

The numerical procedure used to obtain the time-dependent solutions (Equation 4.16 to 4.24) is based on discretizing the model equations spatially, and then integrating them in time, using the LIMEX solver. The initial conditions are obtained from the steady state concentration profiles, and the magnitudes of the variables describing the flow hydrodynamics ($\underline{u_g}$, $\underline{u_s}$ and ϕ) are obtained from the time dependent evolution results based on the linear stability analysis (See Fig 4.19 and 4.20 as examples). All results presented in this section take the dimensionless form:

$$u_x^* = \frac{u_x}{v_t} \qquad C^* = \frac{C}{C_{A0}} \qquad x^* = \frac{x}{\Delta} \qquad y^* = \frac{y}{\Delta}$$

4.4.2 Results and Discussions

To study the effects of flow hydrodynamics on the chain reaction system, the concentration profiles for the steady state are examined as the starting point. The flow profiles are used as the input for the calculation of the concentration profiles are shown in Fig 4.27. Here, two typical flow structures are studied. One represents the "Core-annular" flow, and the other one describes the gas-particle flows with segregation of particles in the center.

When elastic particle-particle collisions are assumed, a "Core-annular" flow can be observed. Based on this flow profile, the variations of the average concentration, average mass flow rate ($<u_x^* C^*>$), and selectivity ($<u_x^* C_B^*>/<u_x^* C_S^*>$) with the bed height are shown in Fig 4.28 and 4.29. Clearly, the concentration of raw material A keeps decreasing when moving upward through the bed, while the concentration of byproduct S keeps increasing. The concentration of product B has a maximum at a certain bed height, and then goes down gradually. The effect of the ratio of the product reaction rate k_A to the byproduct reaction rate k_B on the variations of the concentration profiles with the bed height is significant. When k_A:k_B=1:1, the product concentration C_B has a maximum 0.27 at the bed height 2.2. When k_A:k_B=10:1, C_B increases very quickly, and then passes through a maximum of 0.64 at a bed height of 0.63. The point corresponding to the maximum C_B moves left when the ratio of k_A to k_B changes from 1 to 10 (see Fig 4.28(a) and (b)) because the concentration of raw material C_A consumes much faster when k_A:k_B=10:1 than k_A:k_B=1:1. For k_A:k_B=1:10, the curve representing the consumption of C_A is the same as k_A:k_B=1:1, while the concentration of the product B is much lower in this case. The concentrations of the byproduct C_S are comparable in these three cases. C_S grows the fastest for k_A:k_B=10:1 due to the quick increase in C_B in the initial stage. Although the highest byproduct reaction rate k_B is for the case shown in Fig 4.28(c), the growth of C_S is slower than that in Fig 4.28(b) due to the low level of the product concentration C_B in the third case. A similar mechanism is also shown when plotting the variations of the mass flow rate with the bed height (See Fig 4.29(a) to (c)). In Fig 4.29(d), the selectivity curves always monotonically decrease with an increase in the bed height. It is interesting to note that the increase in the product reaction rate has little effect on the system selectivity while an increase in the byproduct reaction rate tremendously decreases the system selectivity.

When inelastic particle-particle collisions are considered, particles are predicted to segregate near the walls and the flow hydrodynamics exhibits a big difference from those with e_p=1 (See Fig 4.27). Based on this "Core Segregation" flow profile, the variations of the average concentration, average mass flow rate, and selectivity with the bed height are shown in Fig 4.30 and 4.31. By comparing Fig 4.28 with Fig 4.30, and Fig 4.29 with Fig 4.31, it is interesting to note that for a chosen set of flow parameters the variations of the system average properties, such as the average concentration and the average mass flow rate, with the bed height are not sensitive to the variations of the flow profiles. In this chain reaction system, the contributions of the variations of the reaction rates are much more pronounced to the variations of the system average properties.

To further study the effect of the flow hydrodynamics on the chain reaction system, an investigation of the spatial concentration contour map is required. The Concentration contour maps with a "Core Annular" flow structure are shown in Fig 4.32. Clearly, the residence time is short in regions with high gas velocity, and vice versa. Therefore, a variation of the product concentration from a low level to a high level, and then back to a low level is observed at the center, and a high accumulation of the byproduct can be seen near the walls. When particle-particle collisions are inelastic, the gas velocity is low in the regions at the center and near the walls (See Fig 4.27). Concentration contour maps for this case are shown in Fig. 4.33. A high accumulation of byproduct *S* can be observed at the center and near the walls, while most changes in the accumulation and consumption of product *B* are exhibited in the regions between the center and the walls (See Fig 4.33). By comparing Fig 4.32 and Fig 4.33, it is clear that the effect of the flow hydrodynamics on the spatial concentration profiles is significant.

As discussed in the previous section, a gas fluidized bed reactor may lose its stability to small perturbations and the instability may have symmetric patterns or anti-symmetric patterns. It will be of interest to see how these instabilities affect the chain reaction system. Since the gas-particle flow is stable to small perturbations when particle-particle collisions are assumed elastic, we only consider the reaction evolution for the inelastic case.

The temporal evolution of concentration for product B, with anti-symmetric instability patterns is shown in Fig 4.34. At time t=0, the concentration contour map is symmetric between the centerline due to the symmetry of the steady state flow profiles. With increasing time, the concentration profiles start to lose their symmetry due to the snti-symmetric instability. They sway across the bed and the changes become stronger and stronger. Finally, some high concentration regions and low concentration regions can be observed to move through the bed. When t=27.6, a region with a high concentration of B develops in the lower portion of the bed, then moves upward and towards to the left section of the bed (t=28.2). This is due to the instability streamer that sways to the left side leading to a high solids fraction region, or a high catalyst concentration region, in the left section of the bed. A high concentration region can be observed in the flow area before the gas flows out of the bed. With continuously increasing time, a high concentration region develops in the lower portion of the bed again (t=28.8), then it moves upward and towards to the right section of the bed (t=29.4) because the streamer sways to the right side. The high concentration region observed in the concentration profiles corresponds to the solids clusters observed in the flow profiles.

The temporal evolution of concentration for product B with symmetric instability flow patterns is shown in Fig 4.35. With increasing time, the concentration profiles oscillate symmetrically, and the oscillation becomes more and more violent corresponding to the changes of the flow profiles. The evolution of C_B in one periodic time unit is also shown in Fig 4.35. When t=38, a region with high C_B is observed at the lower portion of the bed and two layers with a relative high concentration are symmetric across the centerline. With an increase in time (t=40), the high concentration regions stretch upward and towards the walls like two arms while the region of high concentration at the lower portion of the bed decreases significantly. With an increase in time, a high concentration region starts to develop at the lower portion of the bed again (t=42). With a further increase in time, the variation of the concentration profiles in the bed is increased due to the increase in the variation of the flow profiles (see t=60).

Scrutiny of Fig 4.34 and 4.35 reveals that the effect of instability structures on the concentration profiles is significant. If the instability structure has anti-symmetric patterns, the concentration profile also oscillates anti-symmetrically and the high concentration region generate in the regions of solids clusters. On the other hand, if the instability structure has symmetric patterns, the concentration profile is also symmetric, and C_B has a periodic oscillation between a high concentration state and a low concentration state due to the alternating layers of high density regions and low density regions when the instability traveling waves move through the bed.

4.5 Conclusions

In this chapter, the instability natures of the steady state solutions of gas-particle fluidized beds are investigated by a linear stability analysis approach.

Without considering the effect of boundary conditions, equations of motion have a solution representing a uniform fluidized bed. One dimensional stability analysis shows that the uniform state is stable to lateral perturbations, while unstable to vertical perturbations if the wavelength is larger than a certain value. The vertical instability has a form of traveling waves and the transverse waves are stationary in the bed. When the disturbances are considered in two dimensions, it is found that the origin of the instability is always traveling waves and its development is much faster in the vertical direction than the lateral direction.

With adding walls, the gas-fluidized bed system is much easier to be stabilized due to a much higher level of the granular temperature. Increasing the solids fraction and decreasing the inelasticity can suppress the instability. The dominant instability pattern has two types of structures, symmetric one and anti-symmetric one. With the variations of the solids fraction and the coefficient of particle-particle restitution, transitions exist between these two patterns. For the chosen set of parameters, all eigenmodes are traveling waves, and they move at speeds comparable with that of the solids phase. The instability behaviors are compared between the source wall conditions and sink wall conditions. It is found that both leading eigenvalue profiles and dominant eigenfunction contour maps are insensitive to the physical parameters of the boundary walls due to the similarity of their base state solutions.

The physical mechanism of the instabilities is investigated by a term-by-term analysis. The present study finds that the origin of the instabilities in gas-fluidized beds requires the presence of the inertial terms. If the solids phase inertial term is sufficiently small, all eigenmodes become stable. The gas inertial force and the solid inertial force have different effects on the instability patterns. The development of symmetric patterns is mainly due to the contribution of the gas phase inertial terms, while the development of anti-symmetric patterns is mainly due to the contribution of the solid phase inertial terms. The inelasticity of particle-particle collisions is also necessary for the origin of instability. With reducing the contribution of the dissipation term in Equation 2.5, the instability is rapidly suppressed.

In the present work, a simple chain reaction is used to analyze the effects of flow hydrodynamics on chemical reactions in a gas-particle fluidized bed. For the steady state solutions, the variations of the average properties, such as the average concentration and the average mass flow rate, with the bed height are not sensitive to the variations of the flow profiles, while the contributions of the different reaction rates to the average properties are much more pronounced. The effects of the flow profiles and their instabilities on the reaction profiles, or concentration contour maps are significant. The spatial concentration profiles will oscillate corresponding to the oscillations of the gas-particle flow profiles.

Table 4.1: Basic simulation parameters

d	Particle diameter	400μm
μ_g	Viscosity of gas	0.00004 kg/(ms)
ϕ_{max}	Maximum solid volume fraction	0.65
ρ_g	Gas density	3 kg/m^3
ρ_s	Solid density	1600 kg/m^3
N	Richardson-Zaki number	3
g	Gravitational acceleration	9.81 N/kg
$2\Delta/d$	Ratio of bed width to particle diameter	60
φ'	Specularity coefficient	0.5

Fig 4.1: Variations of the growth rate of the leading mode with the vertical wavenumber under one-dimensional vertical perturbations (e_p=0.95).

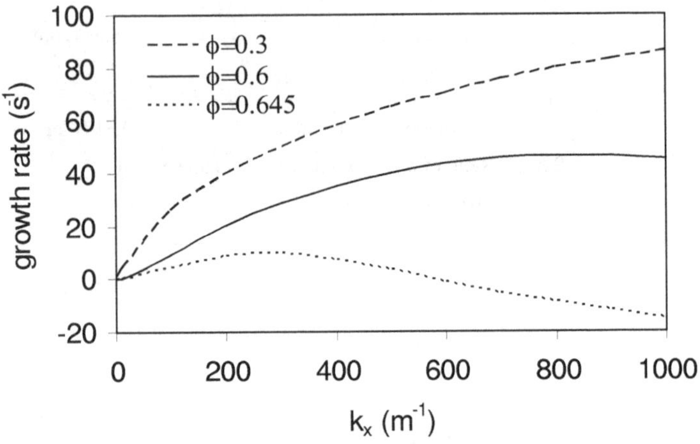

Fig 4.2: a) Variations of the solids phase pressure with the solids fraction. b) Variations of the solids phase effective viscosity with the solids fraction (e_p=0.95).

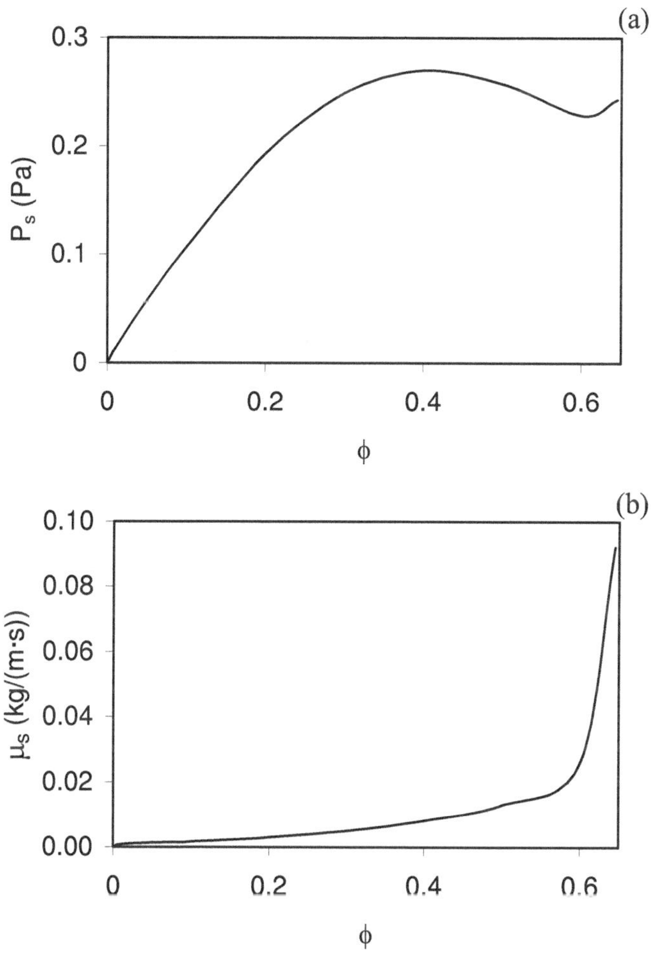

Fig 4.3: Variations of the granular temperature with the solids fraction (e_p=0.95).

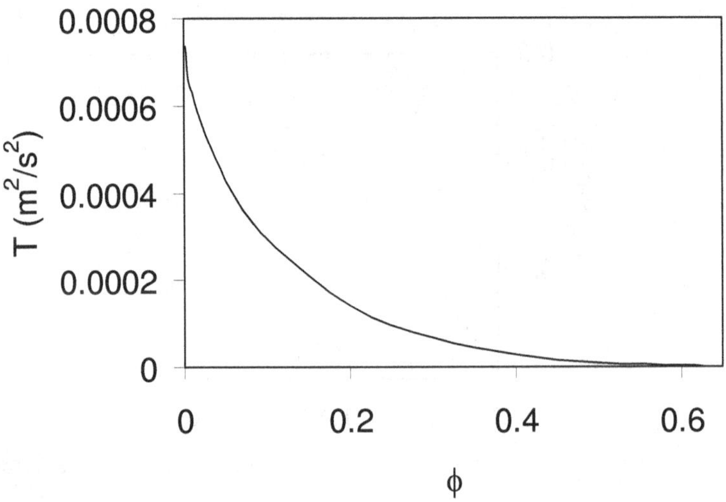

Fig 4.4: Variations of the growth rate of the leading mode with the lateral wavenumber under one-dimensional transverse perturbations (e_p=0.95).

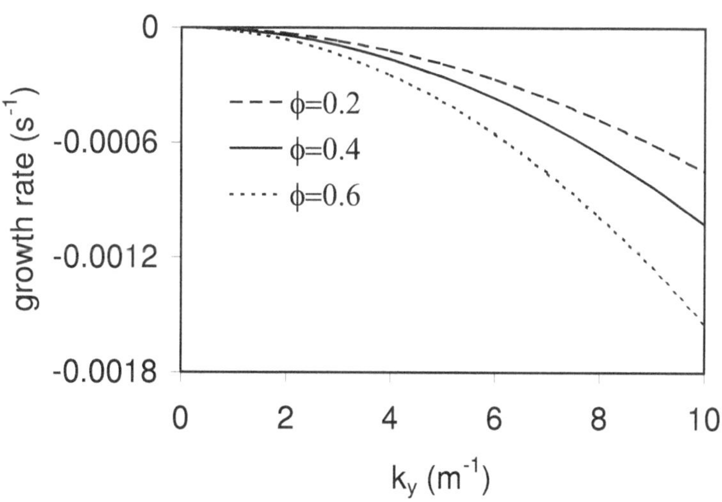

Fig 4.5: Variations of the growth rate of the leading mode and its phase velocity with the lateral wavenumber, where ············ represents k_x =0, ------- represents k_x =1, ——— represents k_x =2, and -·-·-·-·- represents k_x =4 (e_p=0.95, ϕ =0.5). a) growth rate profile. b) phase velocity profile.

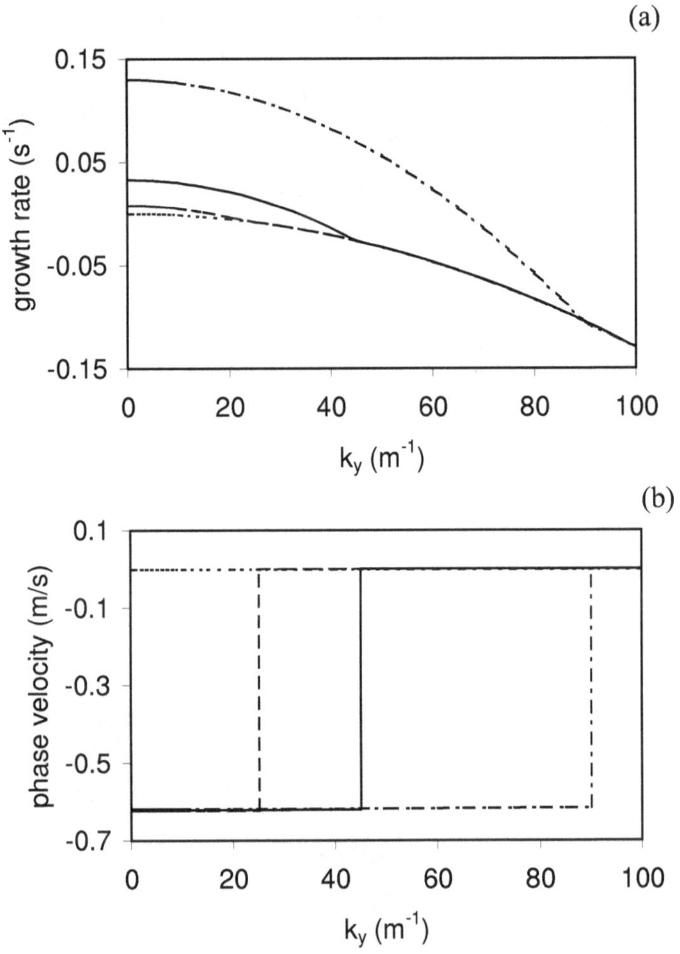

Fig 4.6: Eigenfunctions associated with solids fraction, where k_x =5 and k_y =100 (e_p=0.95, ϕ =0.5).

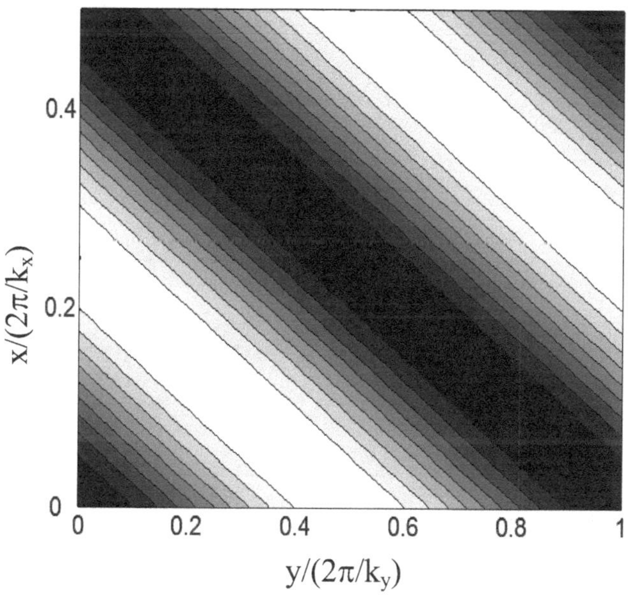

Fig 4.7: Contours of the growth rate of the leading mode in (k_x- k_y) plane (ϕ =0.5). a) e_p=0.9. b) e_p=0.95. c) e_p=1.

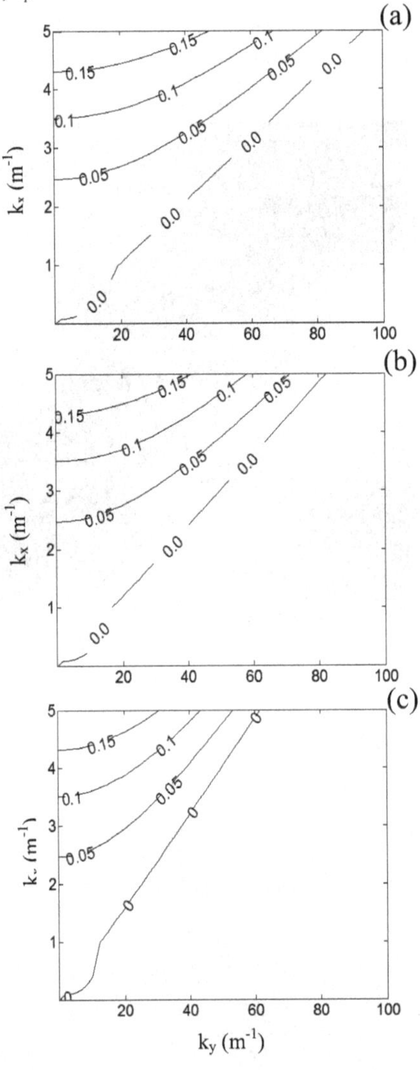

Fig 4.8: Base state profiles for source wall condition (e_p=0.96, e_w=0.98) a) solids fraction profile. b) granular temperature profile. c) gas velocity profile. d) solid velocity profile.

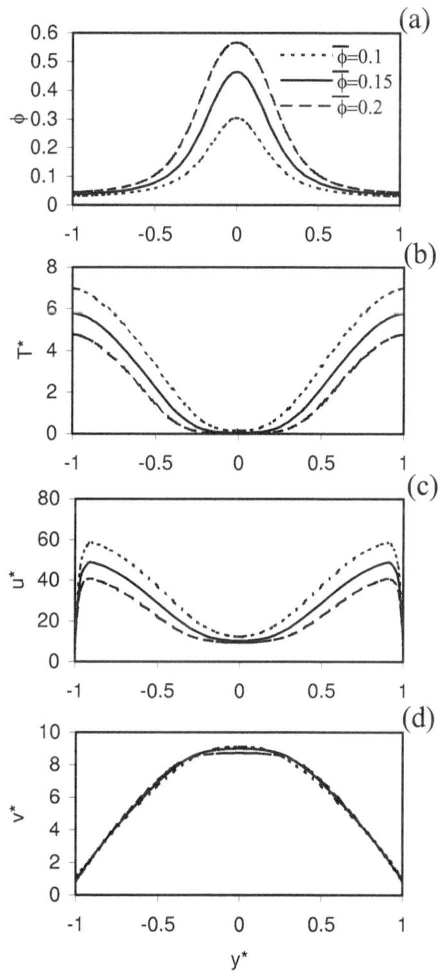

Fig 4.9: Base state profiles for sink wall conditions (e_p=0.96, e_w=0.6) a) solids fraction profile. b) granular temperature profile. c) gas velocity profile. d) solid velocity profile.

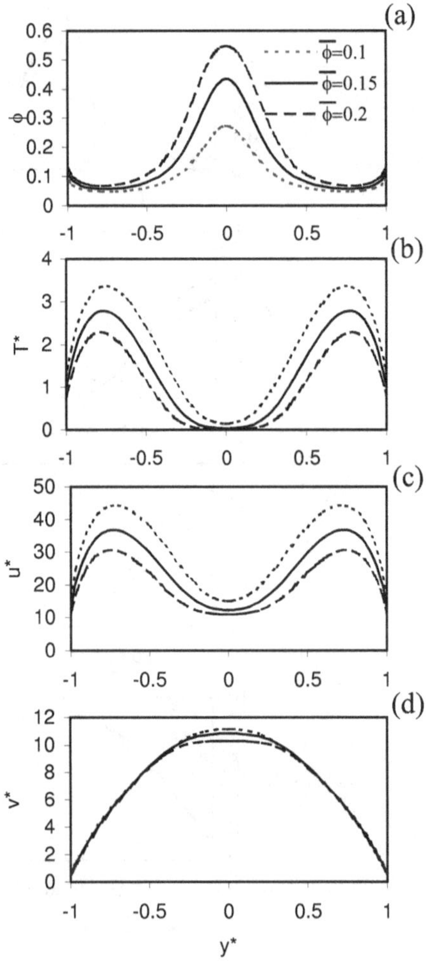

Fig 4.10: Base state profiles for the source wall (e_w=0.99) and sink wall condition (e_w=0.6) when the system is slightly inelastic (e_p=0.99, $\bar{\phi}$ =0.15) a) solids fraction profile. b) granular temperature profile. c) gas velocity profile. d) solid velocity profile.

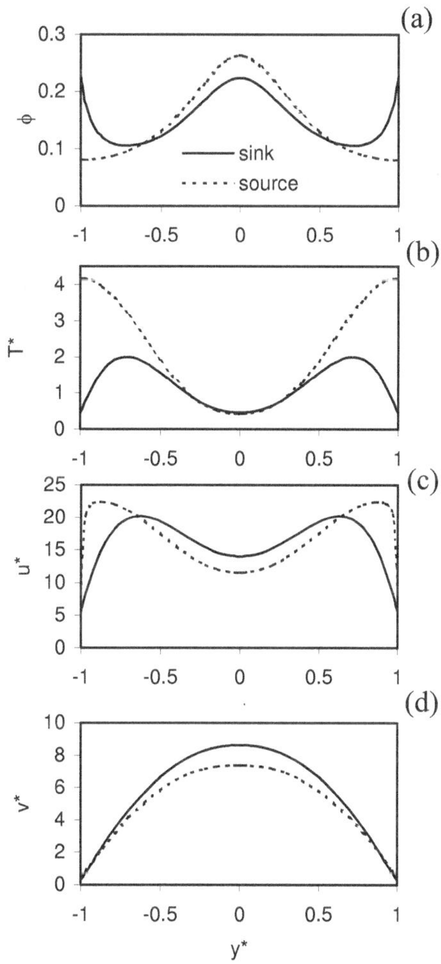

Fig 4.11: Contours of the density patterns of the dominant eigenvalues in (ep-$\bar{\phi}$) plane.

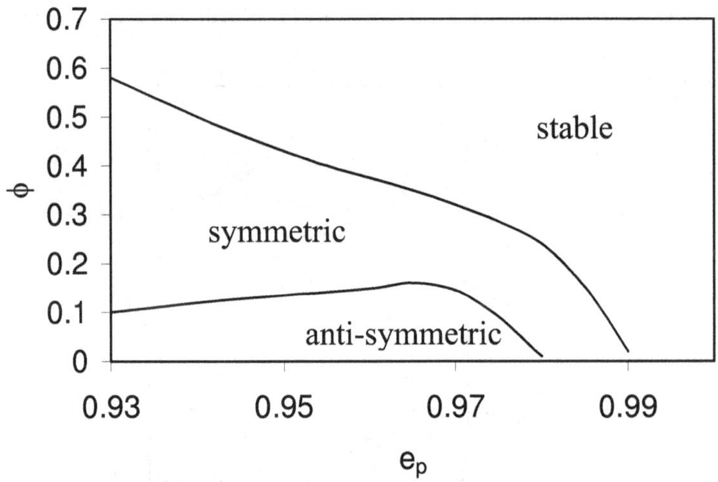

Fig 4.12: Variations of the growth rate of the leading mode and its phase velocity with k_x (e_p=0.96, e_w=0.6), where ··········· represents $\bar{\phi}$ =0.1, ——— represents ϕ =0.15, and ------- represents ϕ =0.2. a) growth rate profile. b) phase velocity profile.

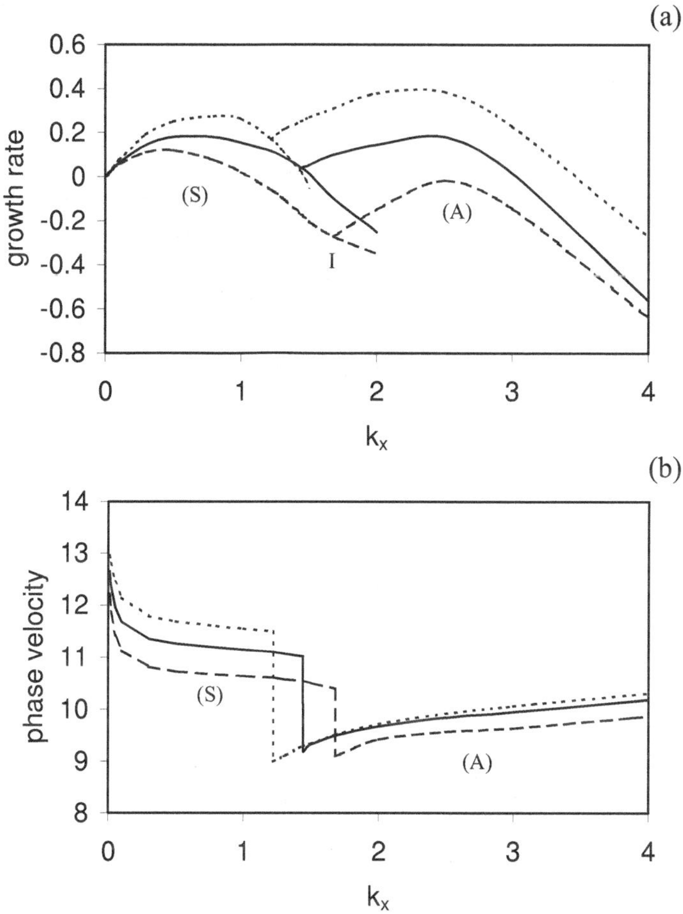

Fig 4.13: Eigenfunctions of the dominant mode with e_p=0.96, e_w=0.6 and $\bar{\phi}$ =0.2. a) Eigenfunctions associated with solids fraction. b) Eigenfunctions associated with solids phase velocity.

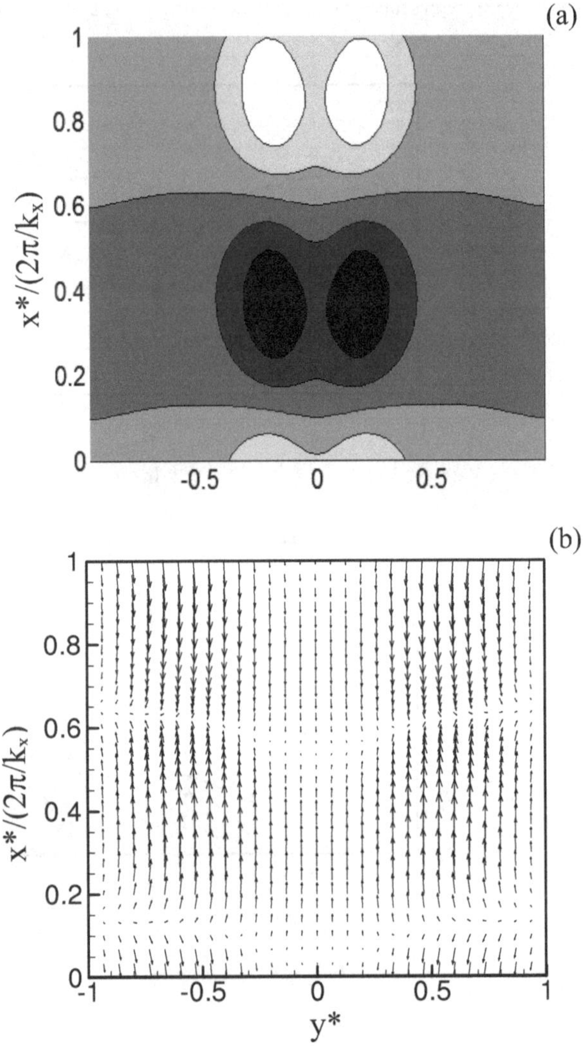

Fig 4.14: Eigenfunctions of the dominant mode with e_p=0.96, e_w=0.6 and $\bar{\phi}$ =0.1. a) Eigenfunctions associated with solids fraction. b) Eigenfunctions associated with solids phase velocity.

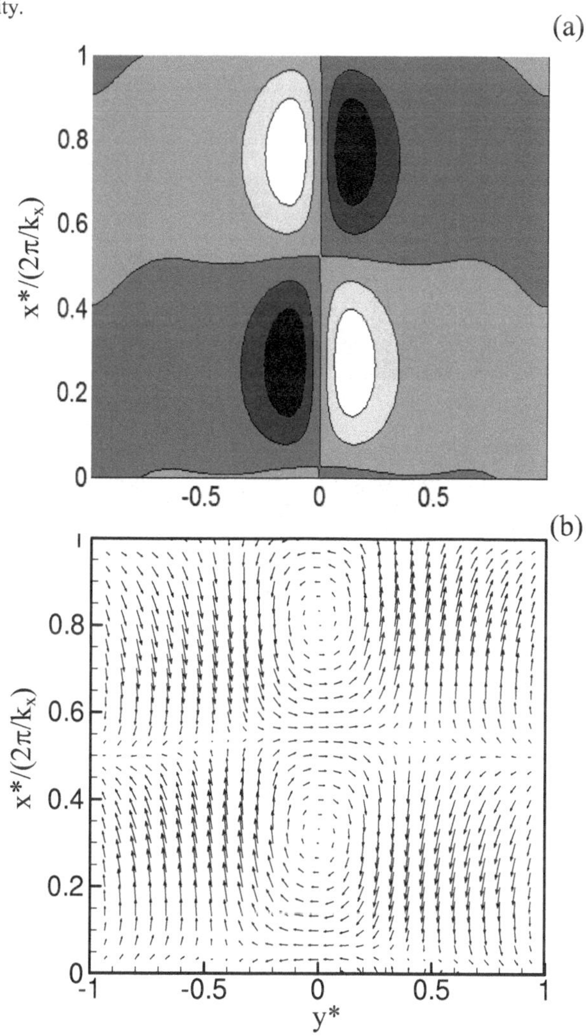

Fig 4.15: Eigenfunctions of the leading mode associated with the solids fraction with e_p=0.96, e_w=0.6 and $\overline{\phi}$ =0.15. a) k_x = 0.65. b) k_x = 1.44(left). c) k_x = 1.44(right). d) k_x = 2.5.

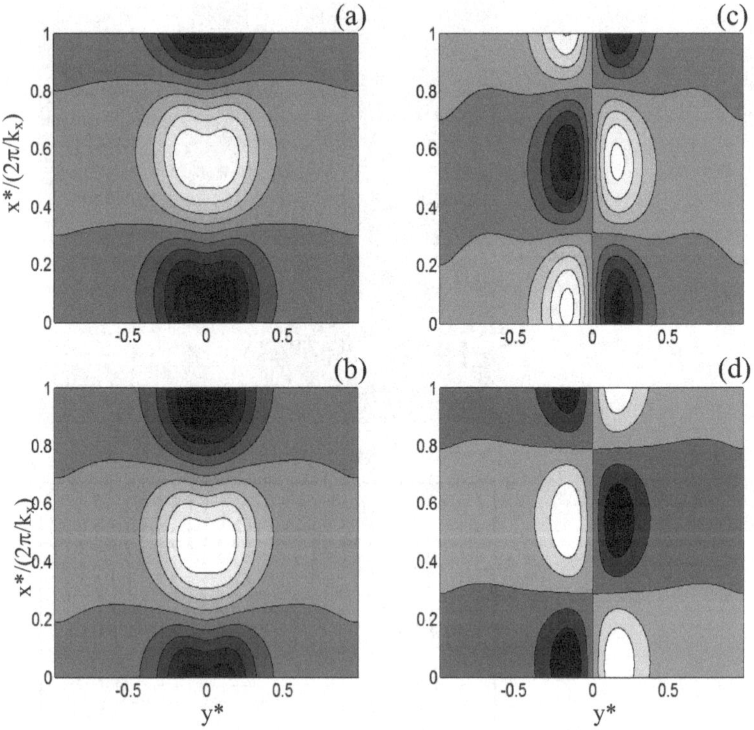

Fig 4.16: Variations of the growth rate of the leading mode and its phase velocity with k_x (e_w=0.6, $\bar{\phi}$ =0.15), where ··········· represents e_p=0.98,——— represents e_p=0.965, and ------- represents e_p=0.95. a) growth rate profile. b) phase velocity profile.

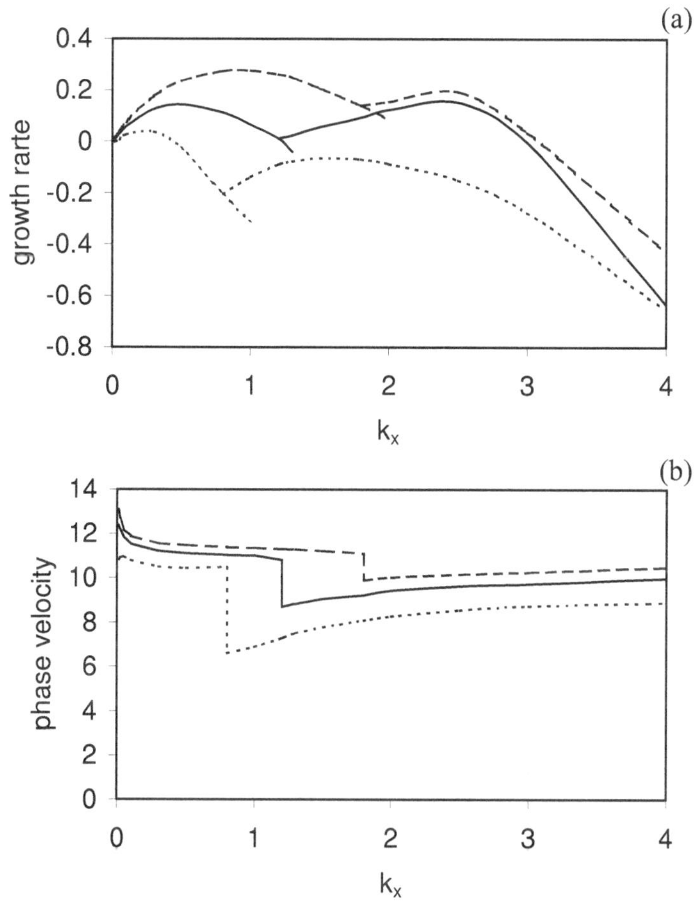

Fig 4.17: Variations of the growth rate of the leading mode and its phase velocity with k_x (e_p=0.96, e_w=0.98), where·················· represents $\bar{\phi}$ =0.2,———— represents $\bar{\phi}$ =0.1. a) growth rate profile. b) phase velocity profile.

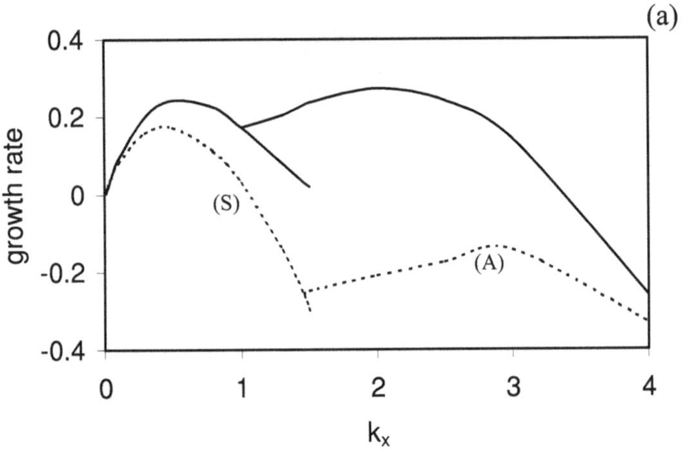

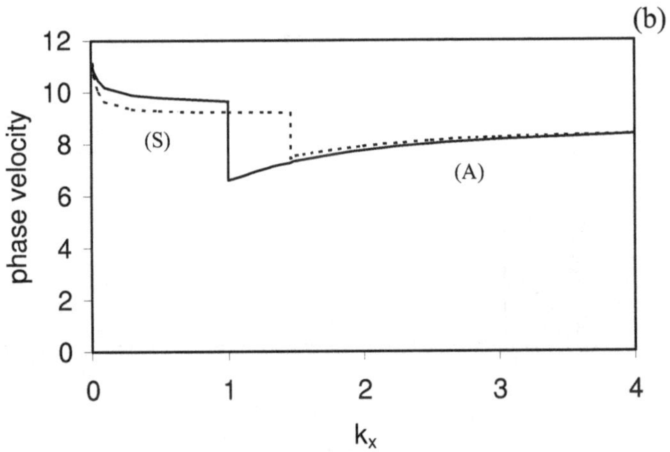

Fig 4.18: Eigenfunctions associated with the solids fraction for the dominant mode with e_p=0.96 and e_w=0.98. a) $\bar{\phi}$ =0.1. b) $\bar{\phi}$ =0.2.

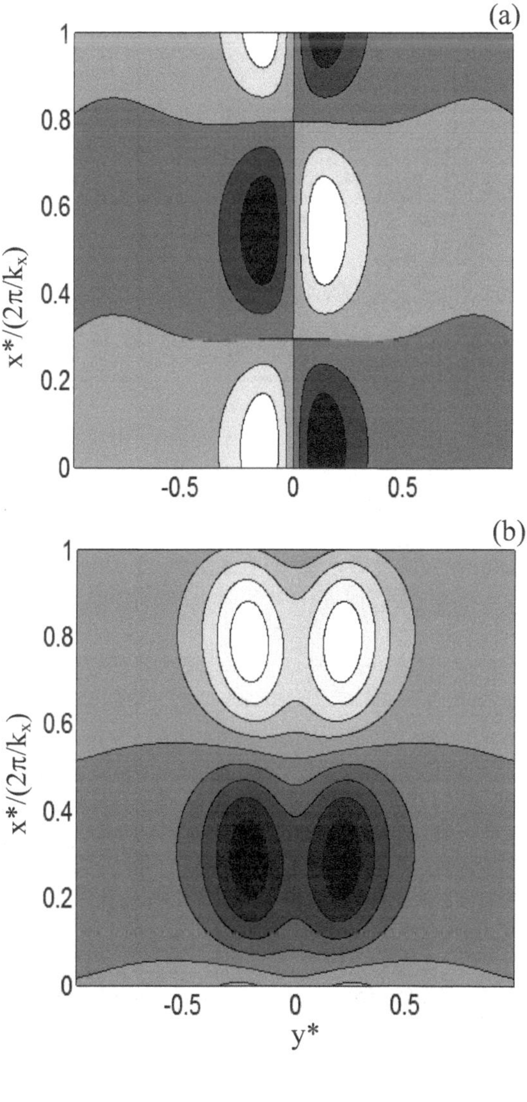

Fig 4.19: Short temporal evolution of density waves with anti-symmetric dominant instability patterns where the scale for the contour plots is from 0 (white) to 0.65 (black). (e_p=0.96, e_w=0.6, $\bar{\phi}$ =0.15, k_x=2.5)

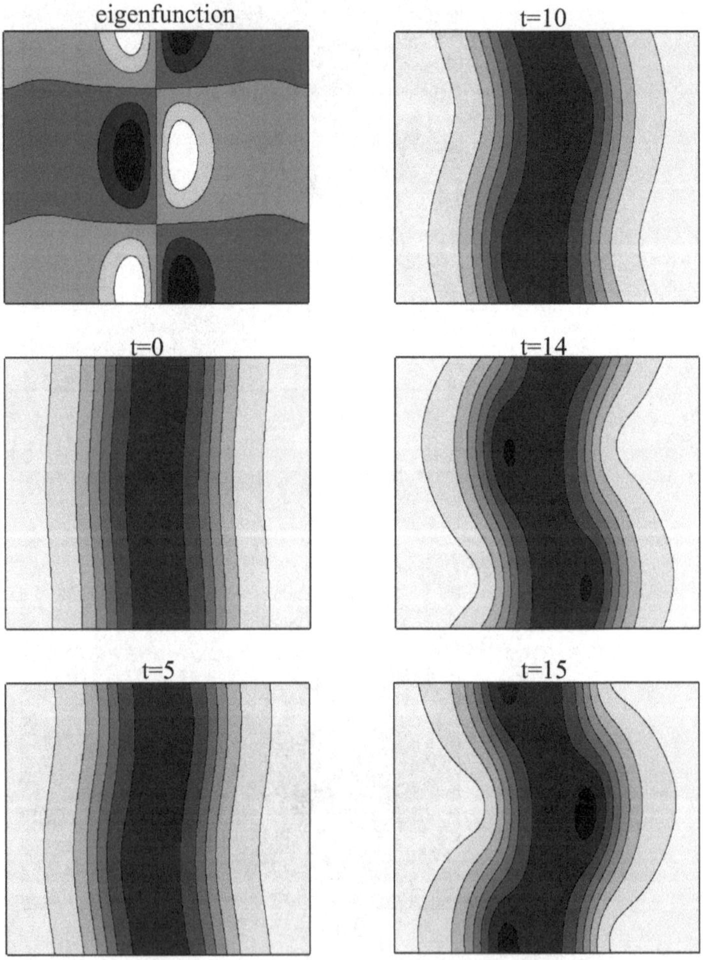

Fig 4.20: Short temporal evolution of density waves with symmetric dominant instability patterns where the scale for the contour plots is from 0 (white) to 0.35 (black). (e_p=0.985, e_w=0.6, $\bar{\phi}$ =0.1, k_x=0.3)

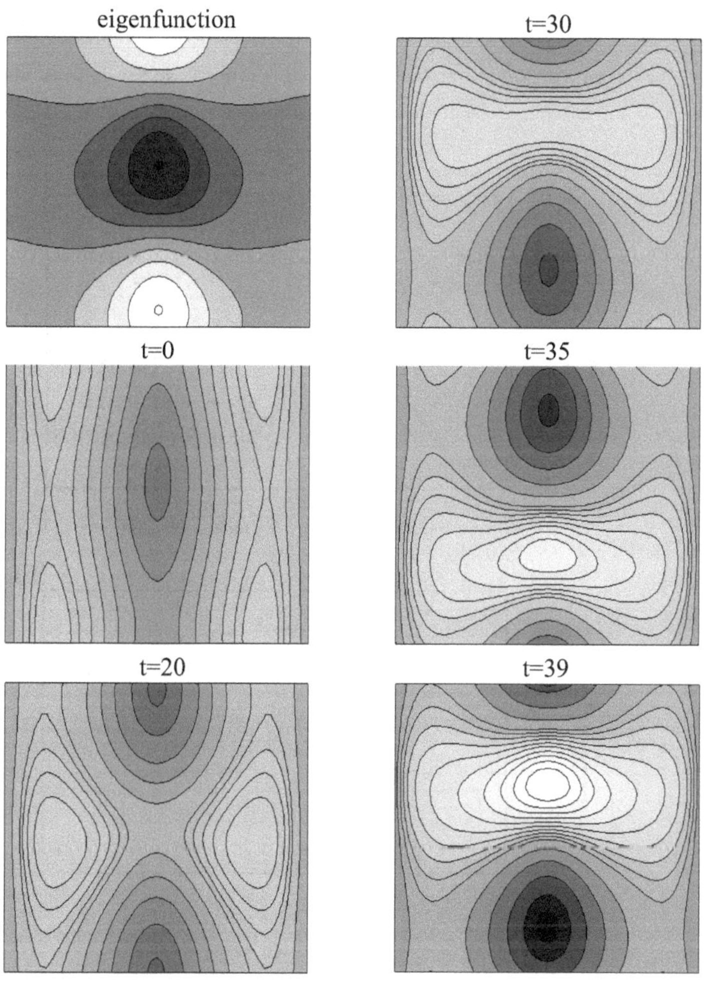

Fig 4.21: Variations of the growth rate of the dominant mode and its phase velocity with the coefficient α (e_p=0.96, e_w=0.6 and $\bar{\phi}$=0.1), where Case I represents the case that α is added onto the inertial term of gas phase, Case II represents the case that α is added onto the inertial term of solid phase, and Case III represents the case that α is added onto the inertial term of both gas and solid phases. a) growth rate profile. b) phase velocity profile.

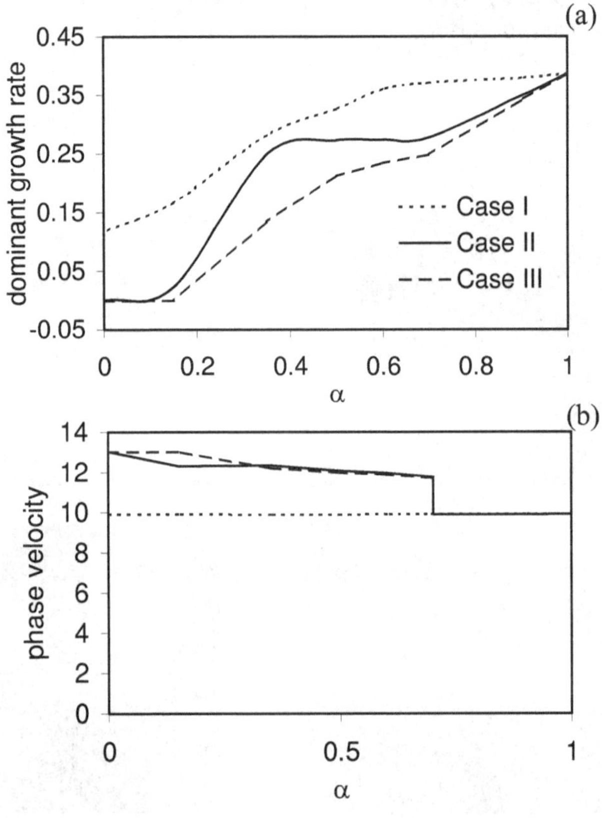

Fig 4.22: Variations of the growth rate of the dominant mode and its phase velocity with the coefficient α (e_p=0.96, e_w=0.6 and $\bar{\phi}$ =0.2), where Case I represents the case that α is added onto the inertial term of gas phase, Case II represents the case that α is added onto the inertial term of solid phase, and Case III represents the case that α is added onto the inertial term of both gas and solid phases. a) growth rate profile. b) phase velocity profile.

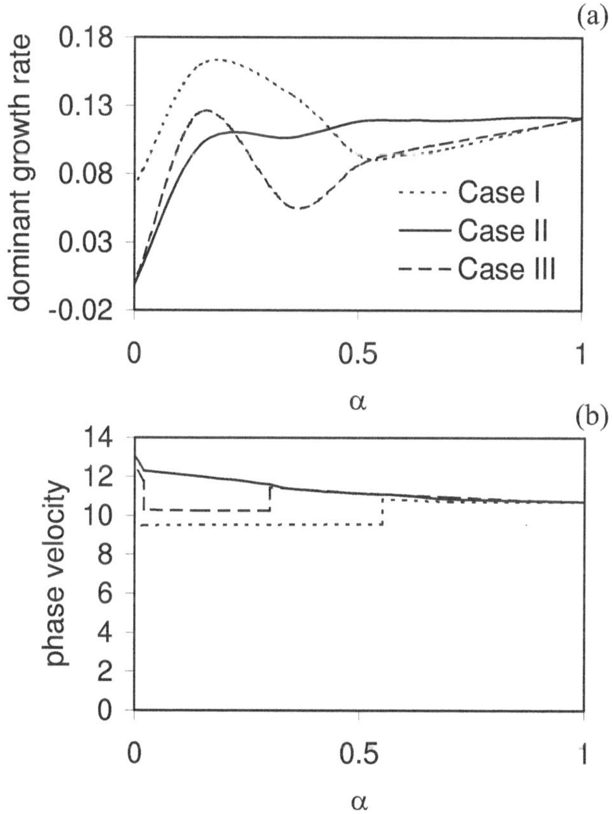

Fig 4.23: Inertial force profiles (e_p=0.96, e_w=0.6). a) based on the symmetric instability patterns ($\bar{\phi}$ =0.2). b) based on the anti-symmetric instability patterns ($\bar{\phi}$ =0.1).

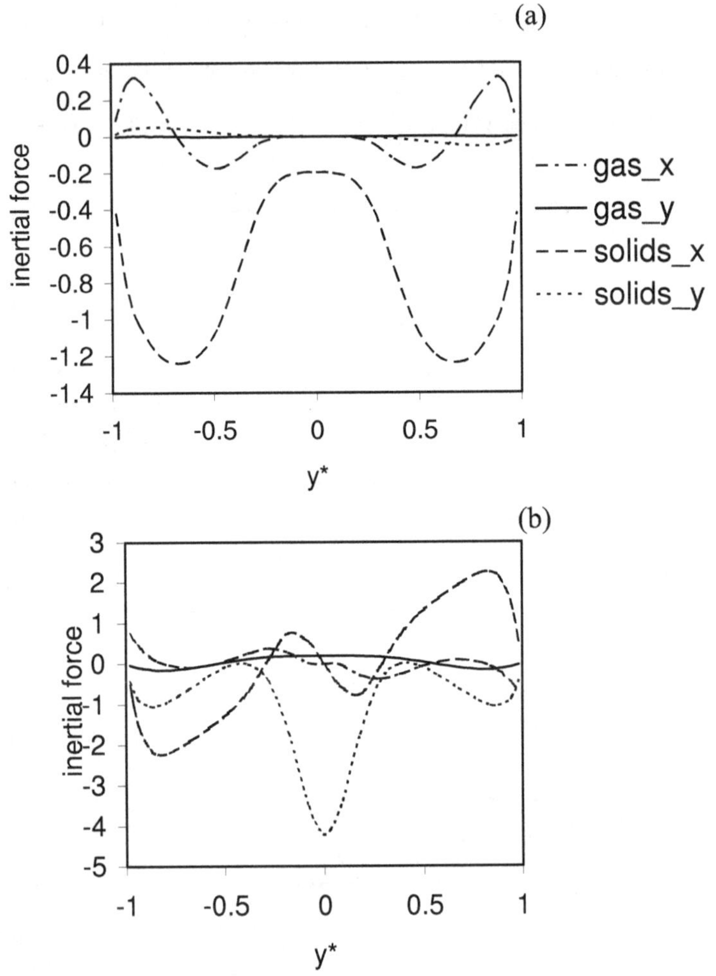

Fig 4.24: Base state profiles when α is added onto the dissipation term (e_p=0.96, e_w=0.6 and $\bar{\phi}$=0.2). a) solids fraction profile. b) granular temperature profile. c) gas velocity profile. d) solid velocity profile.

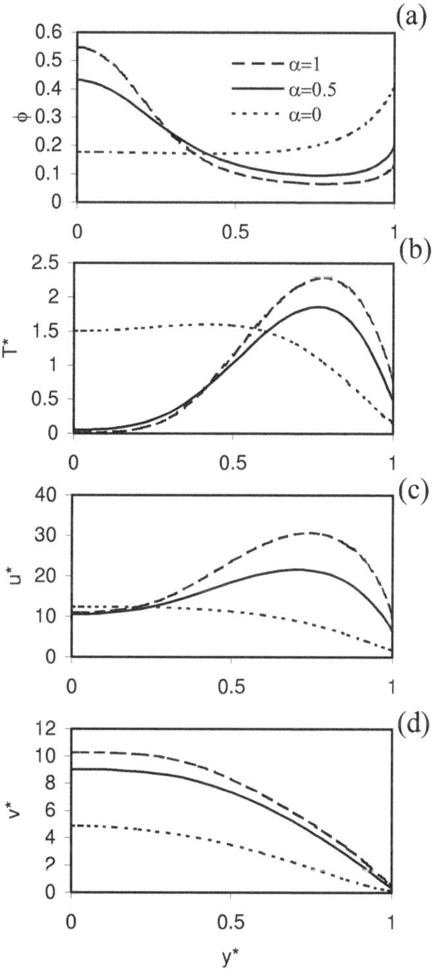

Fig 4.25: Variations of the growth rate of the dominant mode and its phase velocity with the coefficient α when it is added onto the dissipation term (e_p=0.96, e_w=0.6 and $\overline{\phi}$ =0.2). a) growth rate profile. b) phase velocity profile.

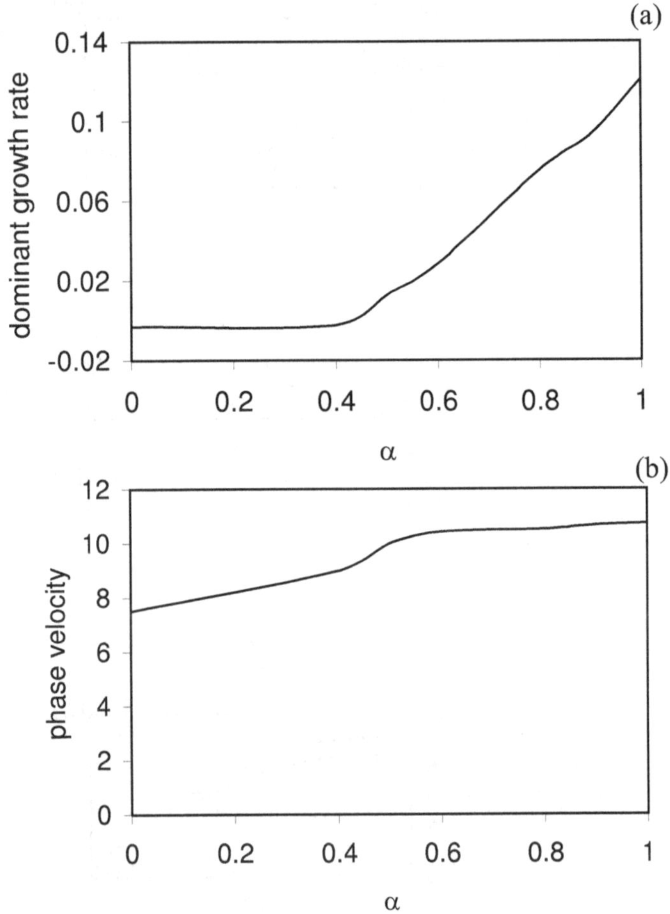

Fig 4.26: Variations of the growth rate of the dominant mode and its phase velocity with the coefficient α when it is added onto the dissipation term (e_p=0.96, e_w=0.6 and $\bar{\phi}$ =0.1). a) growth rate profile. b) phase velocity profile.

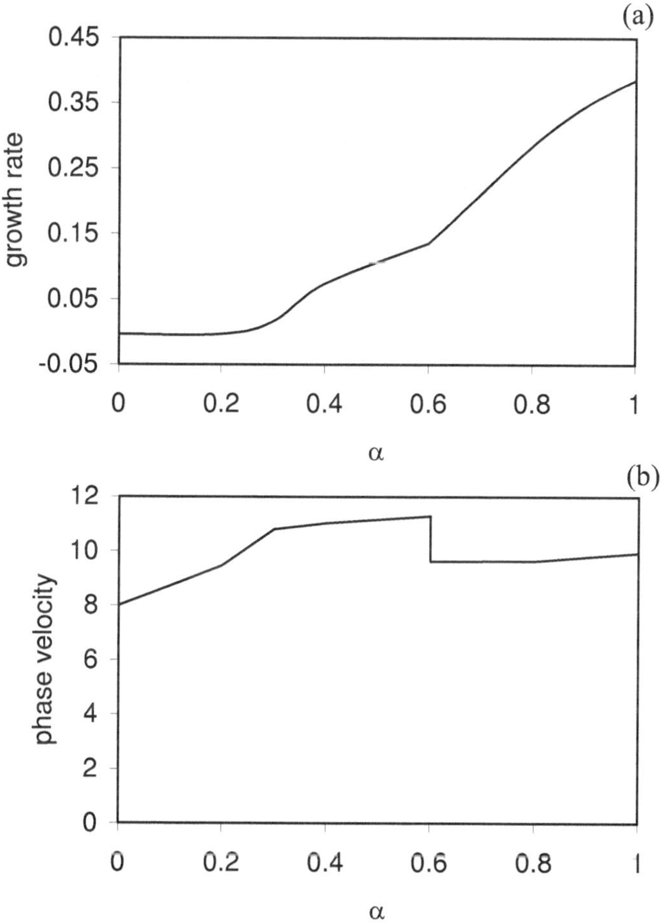

Fig 4.27: Steady state flow profiles for gas-particle fluidized beds ($\bar{\phi}$ =0.1, e_w=0.6). a) solids fraction profile b) granular temperature profile c) gas velocity profile d) solid velocity profile

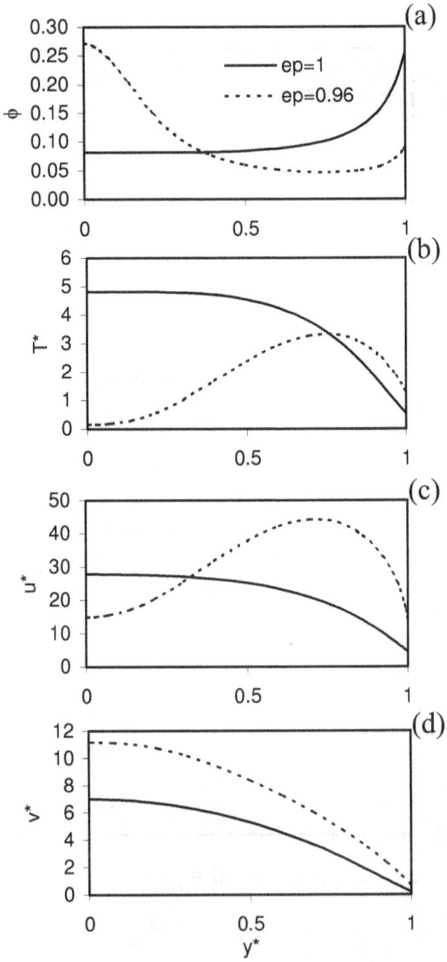

Fig 4.28: Effect of the reaction rate on the variation of concentrations with a "Core Annular" flow structure ($\bar{\phi}$ =0.1, e_p=1, e_w=0.6). a) k_A:k_B=1:1 b) k_A:k_B=10:1 c) k_A:k_B=1:10

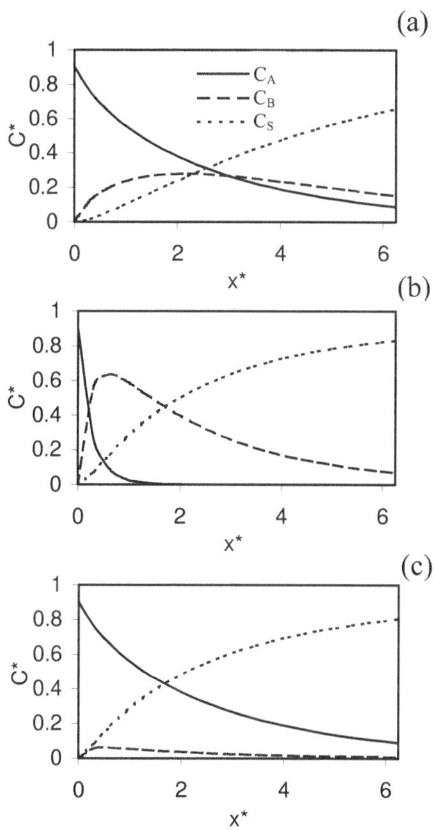

Fig 4.29: Effect of the reaction rate on the yields and selectivity with a "Core Annular" flow structure ($\bar{\phi}$ =0.1, e_p=1, e_w=0.6). a) k_A:k_B=1:1 b) k_A:k_B=10:1 c) k_A:k_B=1:10 d) selectivity profile

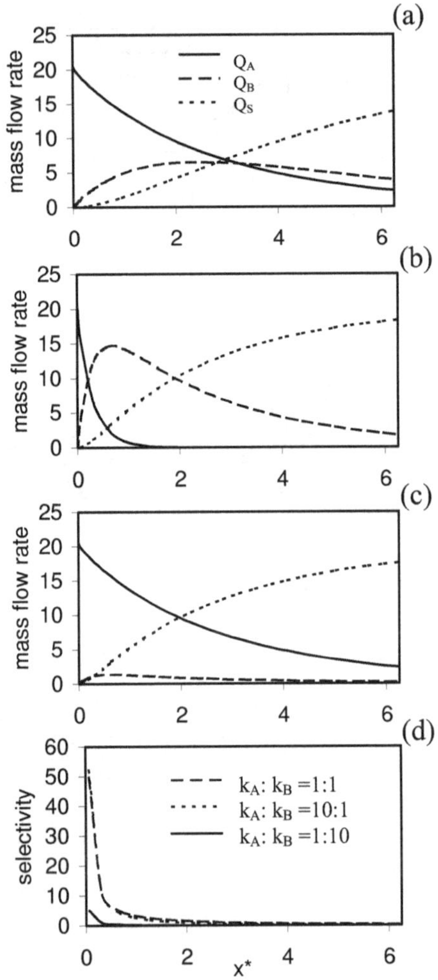

Fig 4.30: Effect of the reaction rate on the variation of concentrations with a "Core Segregation" flow structure ($\bar{\phi}$ =0.1, e_p=0.96, e_w=0.6). a) k_A:k_B=1:1 b) k_A:k_B=10:1 c) k_A:k_B=1:10

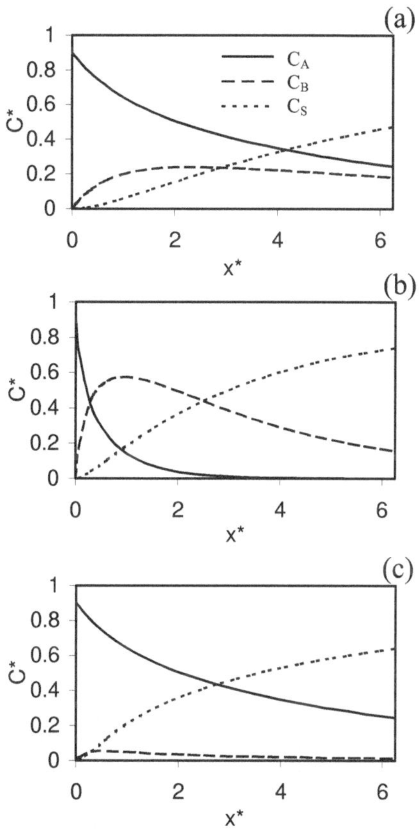

Fig 4.31: Effect of the reaction rate on the yields and selectivity with a "Core Segregation" flow structure ($\bar{\phi}$ =0.1, e_p=0.96, e_w=0.6). a) k_A:k_B=1:1 b) k_A:k_B=10:1 c) k_A:k_B=1:10 d) selectivity profile

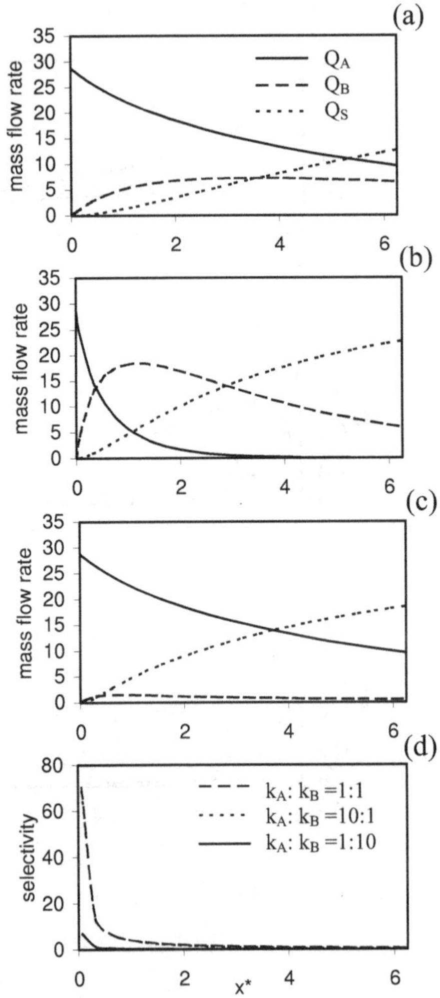

Fig 4.32: Concentration contour maps with a "Core Annular" flow structure where the scale for the contour plots is from 0 (white) to 1 (black) ($\bar{\phi}$ =0.1, e_p=1, e_w=0.6, k_A:k_B=1:1). a) Species *A* b) Species *B* c) Species *S*

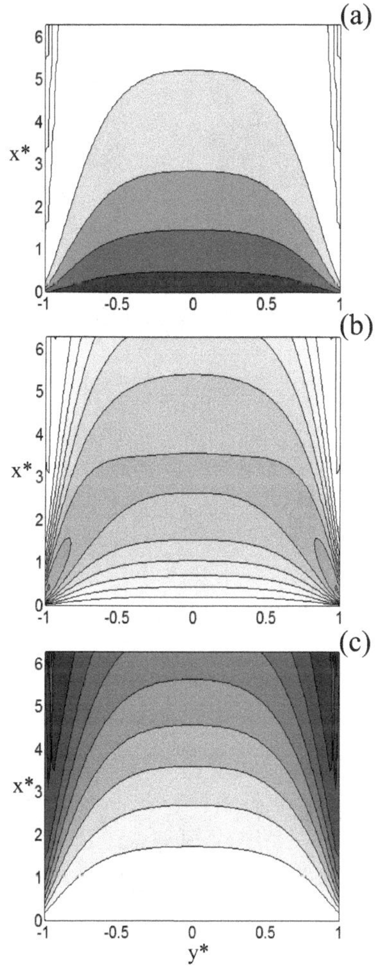

Fig 4.33: Concentration contour maps with a "Core Segregation" flow structure where the scale for the contour plots is from 0 (white) to 1 (black) ($\bar{\phi}$ =0.1, e_p=0.96, e_w=0.6, k_A:k_B=1:1). a) Species A b) Species B c) Species S

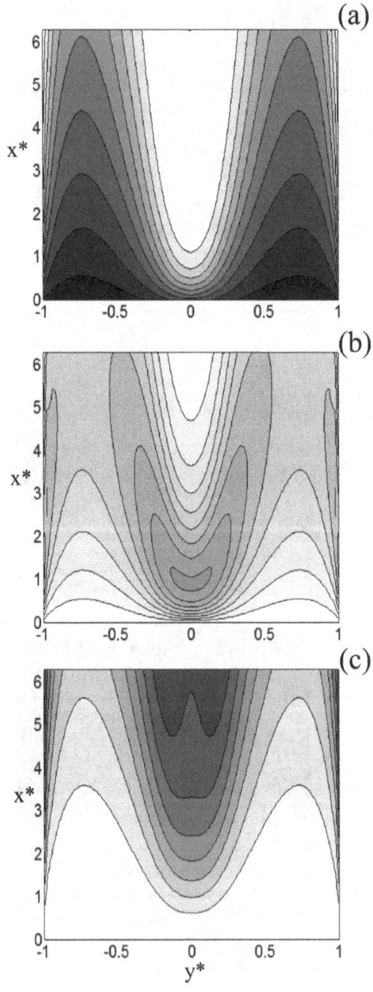

Fig 4.34: Temporal evolution of species *B* concentration contour maps with anti-symmetric instability flow patterns. (e_p=0.96, e_w=0.6, $\bar{\phi}$ =0.1, k_x=2.5, k_A:k_B=1:1) The range of x* is $0<x^*<6$, and y* is $-1<y^*<1$.

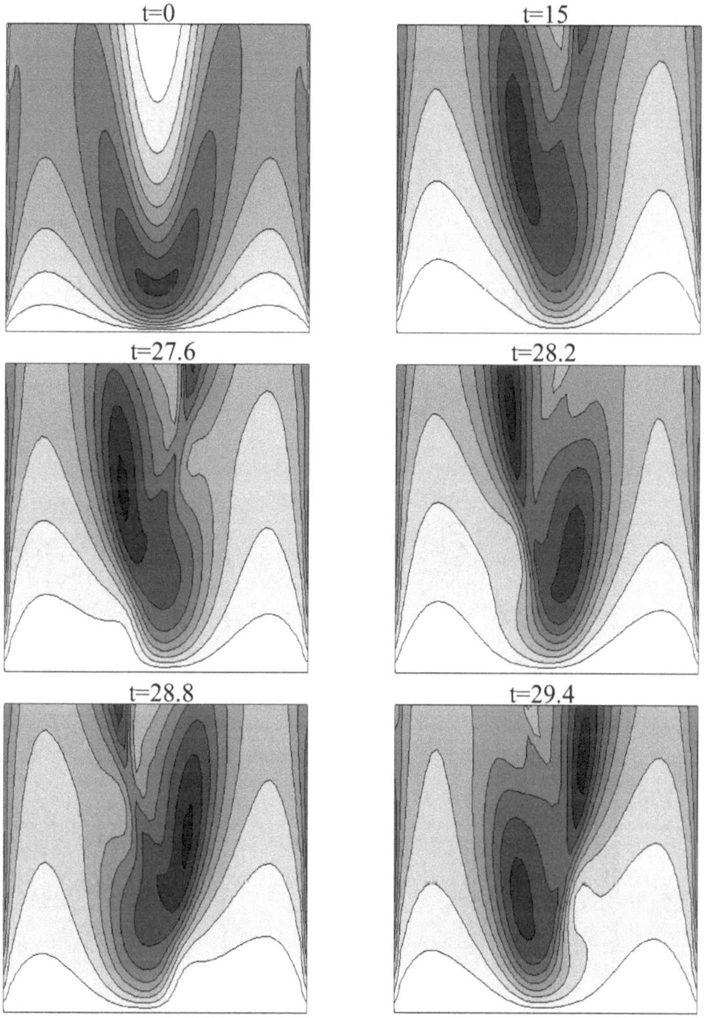

Fig 4.35: Temporal evolution of species *B* concentration contour maps with symmetric instability flow patterns. (e_p=0.96, e_w=0.6, $\bar{\phi}$ =0.1, k_x=1, k_A:k_B=1:1) The range of x* is $0<x^*<6$, and y* is $-1<y^*<1$.

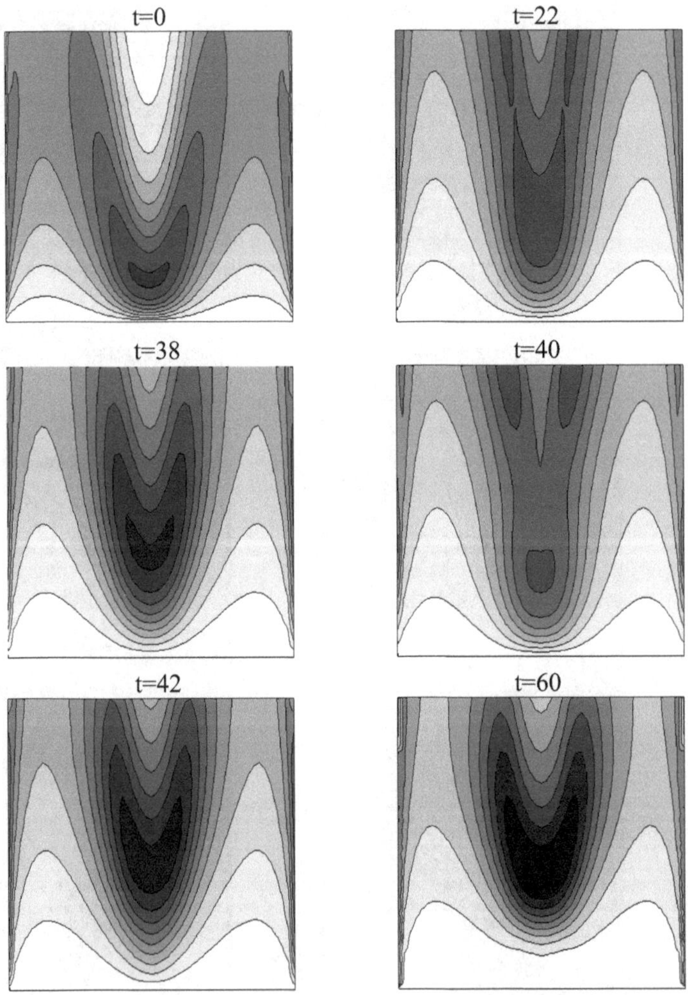

Chapter 5 Steady State Solutions in Granular and Gas-Particle Flows with a Bimodal Particle Mixture

In Chapter 3 and 4, the steady state flow profiles and their stability features in gas-particle fluidized beds with a mono-disperse particle mixture have been examined. However, a real system is always characterized by some degrees of poly-disperse particles that have different sizes and densities. Differences in either size or density of particles may lead to flow-induced segregation, which can affect the bulk behavior of the flowing granular materials and thus the reaction rate and selectivity of fluidized bed reactors. Therefore, it is of interest to extend our investigation to granular and gas-particle flows with a binary particle mixture. To the best of our knowledge, for granular flows, component segregation and non-uniform flow structures have not been reported based on continuum models.

5.1 Model Equations and Numerical Method

In this chapter, three different systems are examined: granular Couette flow, granular channel flow, and gas-particle flow in a channel (a fast fluidized bed).

5.1.1 Granular Couette Flow

In this section, we employ Couette flows as a metaphor for a variety of more complex, equipment specific geometries because Couette flows have proven invaluable in the investigation of dynamics, stability and mixing (Cross and Hohenburg, 1993; Rudolph et al., 1998; Taylor, 1923) of granular flows. They encompass both shear and physical boundary interactions that are essential ingredients of most practical applications. The geometry of the system considered is shown in Fig 5.1, where parallel walls move in equal but opposite velocities and 2Δ is the gap width.

For steady, fully-developed Couette flows, the continuity equations are automatically satisfied and the diffusion velocity $\underline{v_i'}$ is zero (Jenkins and Mancini, 1987). Using m_A, v_0, Δ and Δ / v_0 as characteristic mass, velocity, length and time, the governing equations can be cast in dimensionless form. To study the effect of non-equipartition of granular energy, two models are examined in this work. The first model is based on the work of Jenkins and Mancini (1989) (JM model) with a modification by Aranson and Willits (1998). The momentum and pseudo-thermal energy equations for the steady state flow have the form:

$$\frac{d}{dy^*}(\mu_{JM}^* \frac{dv_x^*}{dy^*}) = 0 \tag{5.1}$$

$$\frac{d}{dy^*}(P_{SA-JM}^*) = \psi_{A-y}^* \tag{5.2}$$

$$\frac{d}{dy^*}(P_{SB-JM}^*) = \psi_{B-y}^* \tag{5.3}$$

$$3\mu_{JM}^* (\frac{dv_x^*}{dy^*})^2 + \frac{d}{dy^*}(q_{JM-y}^*) = D^2 \gamma^* \tag{5.4}$$

where $\mu_{JM}^* = \mu_{A-JM}^* + \mu_{B-JM}^*$ represents the total particle viscosity, and P_{Si-JM}^* represents the solids pressure of particle species *i*. $D = \Delta / d_A$ is the ratio of the half of the gap to the large particle diameter, and a value of 20 is used in the simulations. The dimensionless variables and dimensionless terms are

$$y^* = \frac{y}{\Delta} \qquad v_x^* = \frac{v_x}{v_0} \qquad T^* = \frac{3T}{m_A v_0^2} \qquad D = \frac{\Delta}{d_A} \qquad \mu_{JM}^* = \frac{m_A v_0 \mu_{JM}}{\sqrt{3} d_A^2}$$

$$P_{Si-JM}^* = \frac{m_A v_0^2 P_{Si-JM}}{3 d_A^3} \qquad \underline{\psi_i^*} = \frac{m_A v_0^2 \underline{\psi_i}}{3 d_A^3 \Delta} \qquad \underline{q_{JM}^*} = \frac{m_A v_0^3 \underline{q_{JM}}}{3\sqrt{3} d_A^2 \Delta}$$

$$\gamma^* = \frac{m_A v_0^3 \gamma}{3\sqrt{3} d_A^4} \tag{5.5}$$

The expressions for P_{Si-JM}, $\underline{\psi_i}$, μ_{i-JM}, $\underline{q_{JM}}$ and γ can be found in Appendix 1.

The boundary conditions at the walls for the momentum balance and pseudo thermal energy balance are given by Equation 2.33 and 2.34, where the slip velocity $\underline{v}_{sl}=\underline{v}-\underline{v}_0$. It is found that the governing equations along with the dimensionless boundary conditions admit the following symmetry

$$v_x^*(y) = -v_x^*(-y) \qquad \phi_i(y) = \phi_i(-y) \qquad T^*(y) = T^*(-y)$$

Thus, a relevant boundary condition at $y^*=0$ can be written as

$$v_x^* = 0 \qquad \frac{dT^*}{dy^*} = 0 \tag{5.6}$$

The second model is based on the work of Iddir and Arastoopour (2005) (IA model). For steady state Couette flows, the momentum and pseudo-thermal energy equations take the form:

$$\frac{d}{dy^*}[(\mu_A^* + \mu_B^*)\frac{dv_x^*}{dy^*}] = 0 \tag{5.7}$$

$$\frac{d}{dy^*}(P_{SA}^*) = F_{D1A-y}^* \tag{5.8}$$

$$\frac{d}{dy^*}(P_{SB}^*) = F_{D1B-y}^* \tag{5.9}$$

$$3\mu_A^*(\frac{dv_x^*}{dy^*})^2 + \frac{d}{dy^*}(q_{1A-y}^*) = D^2 N_A^* \tag{5.10}$$

$$3\mu_B^*(\frac{dv_x^*}{dy^*})^2 + \frac{d}{dy^*}(q_{1B-y}^*) = D^2 N_B^* \tag{5.11}$$

where the dimensionless terms are

$$\mu_i^* = \frac{m_A v_0 \mu_i}{\sqrt{3} d_A^2} \qquad P_{Si}^* = \frac{m_A v_0^2 P_{Si}}{3 d_A^3} \qquad \underline{F_{D1i}^*} = \frac{m_A v_0^2 \underline{F_{D1i}}}{3 d_A^3 \Delta}$$

$$\underline{q_{1i}^*} = \frac{m_A v_0^3 \underline{q_{1i}}}{3\sqrt{3} d_A^2 \Delta} \qquad N_i^* = \frac{m_A v_0^3 N_i}{3\sqrt{3} d_A^4} \tag{5.12}$$

The constitutive relations are shown in Appendix 1. The boundary conditions of the IA model are given by Equation 2.35, 2.36.

5.1.2 Granular Channel Flow

In this section, the effect of the gravity force is considered and the steady state solutions of granular flows in a channel are examined. The geometry of the granular channel flow is shown in Fig 5.2, where particles move downward due to the gravity force.

From the analysis of granular Couette flows, we find that the non-equipartition of granular energy has a significant effect on the granular flow profiles and particulate segregation (see Section 5.2.1 for details). Therefore, the IA model is used in the following investigations. Using m_A, $v_0 = \sqrt{g\Delta}$, Δ as characteristic mass, velocity, and length, the momentum and pseudo-thermal energy equations can be cast in dimensionless form:

$$\frac{d}{dy^*}(\mu_i^*) + D^2 F_{D1i-x}^* + (3^{1/2} D\phi_i \rho_i / \rho_A \frac{\pi}{6}) = 0 \tag{5.13}$$

$$\frac{d}{dy^*}(P_{Si}^*) = F_{D2i-y}^* \tag{5.14}$$

$$3\mu_i^* (\frac{dv_{xi}^*}{dy^*}) + \frac{d}{dy^*}(q_{1i-y}^* + \sqrt{3} D q_{2i-y}^*) = D^2 N_i^* - 3D^2 u_{xi}^* F_{D1i-x}^* \tag{5.15}$$

where the dimensionless variables and dimensionless terms can be found in Equation 5.5 and 5.12. The boundary conditions are given by Equation 2.35, 2.36.

5.1.3 Gas-Particle Flow in a Channel

Finally, we extend our work to a gas-particle system to study the effect of the gas phase on the flow behaviors and segregation phenomenon. Using m_A, $v_0 = v_{tA}$ (v_{tA} is the terminal velocity of the large particles), Δ as characteristic mass, velocity, and length, the dimensionless governing equations for the fully developed gas-particle flow have the form:

$$-\frac{BD}{C}\frac{dP_f^*}{dx^*} - \frac{BD}{C}[\beta_A^*(u_x^* - v_{xA}^*) + \beta_B^*(u_x^* - v_{xB}^*)] - \frac{BDE}{C} + \frac{d}{dy}(\mu_{eg}^* \frac{du_x^*}{dy^*}) = 0 \tag{5.16}$$

$$\frac{d}{dy^*}(\mu_i^*) + D^2 F_{D1i-x}^* - 3^{1/2}\frac{D}{C}(\frac{n_i}{d_A^3} - \frac{\phi_i}{\phi_A}\frac{n_A}{d_A^3}E) + 3^{1/2}\frac{D}{C}\frac{n_A}{\phi_A d_a^3} \\ (1 - \phi_A - \phi_B + \phi_i)(u_x^* - v_{xi}^*) = 0 \tag{5.17}$$

$$\frac{d}{dy^*}(P_{Si}^*) = F_{D2i-y}^* \tag{5.18}$$

$$3\mu_i^* (\frac{dv_{xi}^*}{dy^*}) + \frac{d}{dy^*}(q_{1i-y}^* + \sqrt{3} D q_{2i-y}^*) = D^2 N_i^* - 3D^2 u_{xi}^* F_{D1i-x}^* \tag{5.19}$$

where the dimensionless parameters and dimensionless variables are

$$B = \frac{d_A \rho_A v_0}{\mu_g} \quad C = \frac{v_0^2}{g\Delta} \quad E = \frac{\rho_f}{\rho_A} \quad P_f^* = \frac{P_f}{\rho_A g \Delta} \quad u_x^* = \frac{u_x}{v_0} \quad \mu_{eg}^* = \frac{\mu_{eg}}{\mu_g} \tag{5.20}$$

In this model, P_f is the gas phase pressure, and μ_{eg} is the effective gas viscosity. We adopt the expression for μ_{eg} suggested by Sinclair and Jackson (1989), given by Equation 2.7. For the drag coefficient, β_i, an ideal state is assumed (Howley and Glasser, 2002) so that the form, originally proposed by Anderson and Jackson (1968), can be adopted for the bulk of our work.

$$\beta_i = \frac{\phi_i(\rho_i - \rho_f)g(1-\phi)^{1-N}}{v_{ti}} \tag{5.21}$$

where the evaluation of the terminal velocity of particle species i, v_{ti} is from the work of Kunii and Levenspiel (Kunii and Levenspiel, 1991), and a value of 3 is used in the simulations for the exponent N (Richardson and Zaki, 1954) considering the normal flow velocity of gas-fluidized beds.

For the gas-particle system, the boundary conditions at the wall (y* = 1) for the particle phase force balance and pseudo thermal energy balance have the same expression with granular channel flows, as described in Equation 2.35 and 2.36. An appropriate boundary condition for the gas velocity, however, is a subject of debate (see Chapter 2.3) From our previous work of gas particle flows with a monodisperse particle mixture, we have known that the effect of gas boundary conditions on the flow profiles is not significant (see Chapter 3). Therefore, a no-slip gas boundary condition is used in the current work. In the center, the gas phase allows a symmetric boundary condition,

$$u_x^* = 0 \tag{5.22}$$

5.1.4 Numerical Method

The above granular and gas-particle models consist of ordinary differential equations and corresponding unknowns (Granular Couette flow: 4 ODEs and 4 unknowns (v_x^*, ϕ_A, ϕ_B, and T^*) for the JM model, 5 ODEs and 5 unknowns (v_x^*, ϕ_A, ϕ_B, T_A^* and T_B^*) for the IA model. Granular channel flow: 6 ODEs and 6 unknowns (v_{xA}^*, v_{xB}^*, ϕ_A, ϕ_B, T_A^* and T_B^*). Gas-particle channel flow: 7 ODEs and 7 unknowns ($u_x^*, v_{xA}^*, v_{xB}^*$, ϕ_A, ϕ_B, T_A^* and T_B^*)), subject to two point boundary conditions. The solutions are obtained by a numerical method based on orthogonal collocation approach (Constantinides, 1987). 17 collocation points are selected in the simulations to guarantee the convergence and accuracy of the code.

5.2 Results and Discussion

5.2.1 Granular Couette Flow

In this work three different bidisperse systems are considered. (i) a mixture of equal density particles, (ii) a mixture of equal mass particles, and (iii) a mixture of equal size particles. By analyzing these three cases respectively, we can delineate the effect of size disparity (case ii) and the effect of mass disparity (case iii) from the effect of size and mass disparities (case i). The effect of material properties, such as inelasticity, mass ratio and size ratio, on the flow profiles, particle segregation, and the breakdown of equipartition of fluctuational energy are investigated. In the present study two forms of particle segregation are examined. The first one is based on the distribution of the total solids fraction in the system. Thus this is called total solids segregation, which can be described using ϕ_A+ϕ_B. The second one is based on the concentration distribution of each individual particle species. By calculating the difference of the solids fraction between two particle species we can examine the mixing behavior in the solids phase. Therefore, we call this solids species segregation, described by ϕ_A-ϕ_B. In the present study all the results are obtained for mixtures with the equal mean solids fractions of large/heavy and small/light particles ($\chi = \phi_A / \phi_B = 1$).

5.2.1.1 Equal Density Mixture

In this section, we set the particle densities equal to each other (ρ_A=ρ_B), so that their masses vary cubically with the size ratio (m_A/m_B=R^3, R=d_A/d_B). Therefore, the effects of both size and mass disparities are included in this case. The first parameter considered is the coefficient of particle-particle restitution (e_p). In kinetic theory, the energy dissipation in the granular medium is described using this parameter, which leads to the inhomogeneity, segregation, and coherent wave structures in granular flows (Conway and Glasser, 2004). For bidisperse particle mixtures, it has been found that the origin of the breakdown of equipartition of granular energy is determined by this parameter (Alam and Luding, 2003; Barrat and Trizac, 2002). Barrat and Trizac (2002) obtained an expression for the granular energy ratio of a driven granular gas based on kinetic theory, and found the ratio is equal to one if elastic collisions between particles are assumed, and less than one in inelastic systems. Their expression was confirmed by the work of Alam and Luding (2002a). In this study, particles are assumed to have the same inelasticity properties, so we use one e_p to describe the inelasticity of collisions between particles A and A, B and B, and A and B.

The variations of the solids fraction of each particle species, the granular energy and the solids velocity with a variation in e_p are shown in Fig 5.3 using the JM model. Both particle species have a high concentration near the center and a low concentration close to the walls. The granular energy is large in regions where the solids concentration is small, and vice versa. These phenomena have been also observed when modeling monodisperse granular flows (Alam and Nott, 1998; Liu and Glasser, 2006; Wang et al., 1996). With a decrease in e_p the granular energy decreases and the heterogeneity of each particle distribution increases, indicating that the total solids segregation is enhanced. By comparing Fig 5.3(a) and Fig 5.3(b), it can be seen that for each e_p condition the solids fraction profiles of the two particle species are almost identical, indicating a good mixing observed in the solids phase. Therefore, under this dilute situation the JM model predicts that the effect of e_p on the solids species segregation is insignificant, whereas its effect on the total solids segregation across the gap is significant.

Fig 5.4 shows the effect of e_p on the flow profiles based on the IA model. By comparing Fig 5.3 and Fig 5.4, it can be seen that the flow profiles show a qualitative agreement. In both models, the two particle species have a high concentration at the center and a low concentration near the walls, while the granular energies are high near the walls and low at the center. In addition both models predict that with a decrease in e_p the granular energy decreases and the total solids segregation increases. However, the solids fraction distribution of particle species A and B is not equal when using the IA model. A scrutiny Fig 5.4(a) and Fig 5.4(b) reveals that large particles have a slightly higher solids fraction at the center and a slightly lower solids fraction near the walls than small particles, leading to a more heterogeneous distribution of large particles.

To further investigate the effect of e_p on the breakdown of equipartition of granular energy, the variations of the average granular energy with e_p are shown in Fig 5.5(a), where $\overline{T} = \frac{\sum T_k n_k}{\sum n_k}$ (n_k represents the number of particles between the kth collocation point and the $(k+1)^{th}$ collocation point) for the JM model and $\overline{T_i} = \frac{\sum T_{i,k} n_{i,k}}{\sum n_{i,k}}$ (i=A/B) for the IA model. Both models predict that the average granular energy increases with an increase in e_p. The magnitude of the average granular energy predicted from the JM model is lager than that from the IA model. To find how the granular energy is partitioned between the two particle species for the IA model, the variation of the granular energy ratio ($R_T = \overline{T_A} \, / \, \overline{T_B}$) with e_p is plotted in Fig 5.5(b). To investigate the effect of solids inhomogeneity on the energy non-equipartition, the result from a simple unbounded shear flow is also investigated for comparison, using the constitutive relations from the work of Iddir and Arastoopour (2005). For a simple shear flow, a linear streamwise velocity profile along with constant granular energy and constant solids fraction can satisfy the balance equations, and the granular energy can be calculated using the equation

$$\mu_i^* \left(\frac{dv^*}{dy^*}\right)^2 = D^2 N_i^* \qquad \frac{dv^*}{dy^*} = 1 \tag{5.23}$$

Clearly, in both situations the granular energy ratio is equal to one when the collisions of particles are assumed elastic. With a decrease in the restitution coefficient, R_T increases and the increase is more pronounced in the model for Couette flows than that for simple shear flows. When e_p =0.7, R_T =1.84 for the non-simple shear flow and R_T =1.71 for the simple shear flow. The difference in the magnitude of the average granular energies between these two situations is much larger. When e_p =0.9, $\overline{T_A} = 1.10$ and $\overline{T_B} = 0.89$ for the non-simple shear flow, while $\overline{T_A} = 0.23$ and $\overline{T_B} = 0.19$ for the simple shear flow. Therefore, it is clear that under this dilute situation, the heterogeneity of the solids distribution has a great effect on the values of the granular energy whereas it is effect on the breakdown of the equipartition of granular energy is not significant.

The second parameter considered is the size ratio (R). The variations of the flow profiles with R are shown in Fig 5.6 for the JM model. With an increase in R the granular energy decreases, and the solids distributions of both particle species become more heterogeneous, indicating an enhancement of the total solids segregation. The effect of R on the solids species segregation is not significant so that particles A and B are locally at similar concentrations. Using the IA model, the effect of R on the flow profiles is shown in Fig 5.7. At R=2, the ϕ_A and ϕ_B profiles are fairly similar and segregation is not observed to any appreciable degree. For R=3 and R=4, the ϕ_A and ϕ_B profiles start to differ. Using the IA model, for a small size ratio ($R \leq 3$) the concentration of both large particles and small particles monotonically decreases when moving from the center to the walls. When the size ratio increases to 4, however, the solids fraction of small particles increases from the center, passes through a maximum, and then decreases when moving towards the walls. Therefore, a significant segregation is observed at the center, which is not seen in the JM model.

The effect of the size ratio on the average granular energy is shown in Fig 5.8(a) Clearly, the average granular energy decreases with an increase in the size ratio. This is because with increasing the size ratio the number of small particles increases in the system, leading to more inelastic collisions and more energy loss. The value of the granular energy calculated from the JM model is comparable from that calculated from the IA model. The effect of the size ratio on the granular energy ratio can be seen in Fig 5.8(b). When R=1, particle A and particle B are identical, thus $\overline{T_A}/\overline{T_B}$=1. As the size ratio increases, the granular energy ratio increases and the increase in the Couette flow is slightly faster than that for the simple shear flow. The magnitudes of the average granular energies exhibit a significant difference between these two situations. When R=4, $\overline{T_A}=0.1789$ and $\overline{T_B}=0.0286$ for the Couette flow, whereas $\overline{T_A}=0.0460$ and $\overline{T_B}=0.0086$ for the simple shear flow. This is in agreement with the previous results. By comparing Fig 5.5(b) and Fig 5.8(b), it is important to note that although the system inelasticity alone can lead to the breakdown of equipartition of granular energy, the contribution of the size disparity is more significant and it greatly amplifies the non-equipartition of granular energy. In this equal density case, however, the difference in the size ratio automatically leads to a difference in the mass ratio. Therefore, it is not clear whether the size disparity dominates the breakdown of equipartition of granular energy or the mass disparity dominates it. Thus a further investigation of the equal mass and equal size mixtures is required. This will be exhibited in the following sections.

So far the total average solids fraction has been held constant at $\bar{\phi}$=0.02 in the simulations. Under this condition, we found the segregation of total particles and each individual particle species is enhanced when the effect of non-equipartition of granular energy is considered. It is of interest to increase the solids load in the system to see its effect on the flow profiles and segregation phenomena. To quantitatively compare the predictions of the JM model and the IA model, the simulation results are plotted together in Fig 5.9 when the total average solids fraction is held 0.04. It is clear that the total solids segregation is much more pronounced in the IA model than the JM model. We believe this is related to the distribution of the granular energy since it is well known that particles have a potential to segregate in low granular energy regions. We found under our specific situation the granular energy is always higher in the JM model than that in the IA model. A reasonable way for comparing the results is to calculate the ratio of the mixture granular energy in the JM model to that in the IA model. We find this ratio is quite high near the center and decreases from 6.4 at the center to 1.4 at the walls, indicating that the most significant change of the granular energy occurs at the center region. This leads to more particles to migrate to the center in the IA model than the JM model, resulting in a much more pronounced total solids segregation observed in the IA model. The solids velocity is higher in the JM model than the IA model in the region near the walls, and a reverse situation arises near the center, leading to a more nonlinear velocity profile calculated in the IA model.

In Fig 5.9(a), pronounced species segregation can be seen in the IA model, while the solids species segregation predicted by the JM model is small. To understand the physical mechanism leading to the solids species segregation, an investigation of the balance of the diffusion forces is carried out. In kinetic theory, the diffusion equation has the following form,

$$\underline{v_A'} - \underline{v_B'} = -\frac{(n_A + n_B)^2}{n_A n_B} D_{AB} \underline{d_1} \qquad (5.24)$$

where D_{AB} represents the coefficient of ordinary diffusion; $\underline{d_1}$ denotes the diffusion force, which is composed of three contributions, the pressure diffusion force, the thermal diffusion force and the ordinary diffusion force. For steady Couette flows, we assume $\underline{v}_A' - \underline{v}_B' = \underline{0}$ (Jenkins and Mancini, 1987) leading to

$$\underline{d_1} = \underline{0} \tag{5.25}$$

Therefore, the solids species segregation is due to the competition of the pressure diffusion force, the thermal diffusion force and the ordinary diffusion force. For the JM model, D_{AB} and $\underline{d_1}$ have the form:

$$D_{AB} = \frac{3}{(n_A + n_B) g_{AB} (1+e_p) r_{AB}^2} \left(\frac{T m_{AB}}{32 \pi m_A m_B}\right)^{1/2} \tag{5.26}$$

$$\begin{aligned} \underline{d_1} = & \frac{1}{(n_A m_A + n_B m_B)(n_A + n_B) T} \{(n_B m_B \nabla P_{sA-JM} - n_A m_A \nabla P_{sB-JM}) - \frac{\pi}{3} g_{AB} r_{AB}^3 \\ & n_A n_B (1+e_p) T (n_A m_A + n_B m_B) [(\frac{\nabla n_A}{n_A} - \frac{\nabla n_B}{n_B}) + \frac{m_B - m_A}{m_{AB}} \frac{\nabla T}{T}]\} \end{aligned} \tag{5.27}$$

where $m_{AB} = m_A + m_B$ and $r_{AB} = (d_A + d_B)/2$. In Equation 5.27, the term associated with ∇P_{sA-JM} and ∇P_{sB-JM} represents the pressure diffusion force, the term associated with ∇n_A and ∇n_B represents the ordinary diffusion force, and the term associated with ∇T is the thermal diffusion force. For the IA model, the constitutive relations of D_{AB} and $\underline{d_1}$ can be written as,

$$\begin{aligned} D_{AB} = & \frac{(m_A T_A + m_B T_B)^{4.5} n_A n_B (m_A m_B)^{-0.5}}{(n_A + n_B)(m_A + m_B)^4 (T_A T_B)^2 (2\pi)^{0.5} g_{AB} (1+e_p) r_{AB}^2 n_A n_B} \times \\ & \frac{1}{1 + 6 \frac{m_A m_B (T_A - T_B)^2}{(m_A T_A + m_B T_B)^2}} \end{aligned} \tag{5.28}$$

$$\begin{aligned} \underline{d_1} = & \frac{m_A + m_B}{(n_A m_A + n_B m_B)(n_A + n_B)(m T_A + m_B T_B)} [(n_B m_B \nabla P_{sA} - n_A m_A \nabla P_{sB}) \\ & - (n_B m_B + n_A m_A) C_n (\frac{\nabla n_A}{n_A} - \frac{\nabla n_B}{n_B}) - (C_{TA} \nabla T_A - C_{TB} \nabla T_B)] \end{aligned} \tag{5.29}$$

where the thermal diffusion force is associated with ∇T_A and ∇T_B, and the coefficients C_n, C_{TA} and C_{TB} can be found in Appendix 1. The dimensionless expressions of the ordinary diffusion coefficient and the diffusion force can be written as

$$D_{AB}^* = \frac{D_{AB}}{d_A v_0} \qquad \underline{d_1^*} = \Delta \underline{d_1}$$

The profiles of D_{AB}^* and the dimensionless diffusion forces are shown in Fig 5.10 corresponding to the flow profiles in Fig 5.9. Since D_{AB}^* decreases with the solids fraction and increases with an increase in the granular energy, both the JM model and the IA model predict a monotonic increase in D_{AB}^* from the center to the walls corresponding to the profiles of the solids fraction and the granular energy in Fig 5.9. The value of D_{AB}^* is higher in the JM model than the IA model near the center, and a reverse situation arises close to the walls. The profiles of the three diffusion forces calculated for the JM model are shown in Fig 5.10(b). Due to the symmetric boundary conditions, all three diffusion forces have zero values at the center. Then from y*=0 to 1, the ordinary diffusion force and the thermal diffusion force increase quickly from the center, pass through a maximum and decrease when moving towards the walls. The pressure diffusion force has negative values in this range. Therefore, it is clear that the combination of the ordinary diffusion force and the thermal diffusion force is compensated by the pressure diffusion force. Similar results were obtained from particle dynamic simulations for a different geometry (Galvin et al., 2005). When using the IA model, although the profiles are qualitatively similar with the JM model, the absolute values of the diffusion forces are much larger in the IA model than those in the JM model. In addition, in the JM model the contribution of the ordinary diffusion force is much smaller than the thermal diffusion force so that the diffusion is dominated by the balance of the pressure diffusion force and the thermal diffusion force. In the IA model, however, the magnitude of the ordinary diffusion force is much larger and it is almost equal to the thermal diffusion force. From Equation 5.27 and 5.29 we can see that the ordinary diffusion force is proportional to -$(\frac{\nabla n_A}{n_A} - \frac{\nabla n_B}{n_B})$, which describes the difference of the particle concentration gradient between the two particle species. The solids species segregation increases with an increase in the magnitude of the ordinary diffusion force. If this force is zero there is no solids species segregation in the system. Therefore, comparing with the JM model the larger ordinary diffusion force predicted in the IA model leads to a more pronounced solids species segregation calculated from the IA model.

To further understand the physical mechanism leading to the solids species segregation, we rewrite Equation 5.27 and 5.29 based on the work of Kincaid et al. (1987), and Arnarson and Willits (1998). Recalling that the particle pressure is a function of the solids fraction and the granular energy, Equation 5.27 can be rewritten as

$$-\alpha_{AB}\frac{\nabla T}{T} = \frac{\nabla \phi_A}{\phi_A} - \frac{\nabla \phi_B}{\phi_B} \tag{5.30}$$

where the term on the left hand side represents the thermal segregation, and α_{AB} represents the thermal diffusion factor, which can be understood as follows: considering a binary mixture of particles held between two layers which have different granular energies, if $\alpha_{AB}>0$, particle species *A* is more concentrated near the cold layer than species *B*, and species *B* is more concentrated near the hot layer than species *A*. The situation reverses when $\alpha_{AB}<0$. In a similar way, Equation 5.29 can be rewritten as

$$-\alpha_A \frac{\nabla T_A}{T_A} + \alpha_B \frac{\nabla T_B}{T_B} = \frac{\nabla \phi_A}{\phi_A} - \frac{\nabla \phi_B}{\phi_B} \tag{5.31}$$

where α_A (α_B) is the thermal diffusion factor for species *A* (*B*) and the first (second) term on the left side represents thermal segregation for species *A* (*B*).

The profiles of the thermal factors and the thermal segregation are shown in Fig 5.11 corresponding to the flow profiles in Fig 5.9. Using the JM model, the thermal factor has quite small positive values across the walls, indicating that the solids fraction gradient is almost equal between large and small particles. In this case the solids species segregation is not significant. In the IA model, however, the magnitude of the thermal factors is much larger than that in the JM model. The thermal factors of both large and small particles have positive values across the gap, indicating that thermal diffusion drives large particles to move towards the center where the granular energy of large particles is low, and thermal diffusion also drives small particles to move towards the center since the granular energy of small particles is also low at the center. The solids species segregation can be attributed to the additive nature of these two contributions. The thermal segregation term for the IA model (see Fig 5.11b) starts from the zero at the center, decreases rapidly near the center, and then comes back to a value close to zero and plateaus out. This results in the significant solids species segregation observed at the center and the nearly flat solids fraction profiles near the walls (see Fig 5.9). The negative values of the thermal segregation term near the center indicate that large particles have a larger driving force to move towards the center than small particles, leading to the solids fraction of large particles being higher at the center. The variation of the thermal segregation term for the JM model is quite small.

Although in the precious results the solids species segregation is not captured using the JM model (see Fig 5.9), when we increase the total average solids fraction from 0.04 to 0.06, significant solids species segregation can be observed. In Fig 5.12(a), the solids fraction of large particles is much higher than that of small particles near the center, and the situation reverses when moving towards the walls where the profiles reach plateaus. Comparing Fig 5.9(a) (JM model) and Fig 5.12(a) shows that both the total solids segregation and the solids species segregation are enhanced with an increase in the average solids fraction. Similar results are obtained when using the IA model (not shown). The profiles of the three diffusion forces are shown in Fig 5.12(b). By comparing Fig 5.10(b) and Fig 5.12(b), it is clear that the average solids fraction has a significant effect on the diffusion forces. All three diffusion forces (absolute values) increase with an increase in the average solids fraction, and the magnitude of the ordinary diffusion force is larger than the thermal diffusion force. In Fig 5.12(c), the thermal diffusion factor exhibits a big jump at the center and decreases when moving towards the walls, leading to a significant solids species segregation near the center and good mixing of the two particle species observed near the walls. It is interesting to note that the solids species segregation is similar between the JM model with 0.06 average solids fraction and the IA model with 0.04 average solids fraction. This is due to the similarity of the mixture granular energy magnitudes at the center (T_{JM}=0.06 and $T_{IA\text{-}mixture}$=0.05), and the comparable profiles of the diffusion forces.

5.2.1.2 Equal Mass Mixture

In this section, we set the particle mass equal to each other (m_A=m_B), and investigate the effect of the size ratio (R) on the granular flow behaviors. The flow profiles of solids fractions, granular energies and solids velocities for R=4 are shown in Fig 5.13. It can be seen that for the equal mass case the predictions of the JM model and IA model show a good agreement. We believe this is due to the similarity of the granular energy profiles (see Fig 5.13(b)). As was seen for the equal density case, there is total solids segregation in the system where high solids fraction is observed at the center and low solids fraction can be seen near the walls. The solids species segregation, however, is quite different from the equal density case. In the equal density case, the distribution of the large particles is more non-uniform than the small particles, especially in the region near the center where the solids fraction of large particles is much higher than that of small particles (see Fig 5.9(a) for the IA model). In the equal mass case, the situation reverses. Fig 5.13(a) shows that small particles have a high solids fraction at the center and their concentration monotonically decreases towards the walls, while the solids fraction of large particles has a low level at the center, increases and passes through a maximum, and then decreases when moving towards the walls. Therefore, significant solids species segregation can be observed near the center in which the solids fraction of small particles is much higher than large particles.

The variations of the diffusion forces for the JM model and IA model are shown in Fig 5.14(a) and (b), corresponding to the flow profiles in Fig 5.13. For the equal mass case, the pressure diffusion force has positive values from y*=0 to 1 whereas the ordinary diffusion force has negative values. The magnitude of the thermal diffusion force is equal to zero in the JM model and is quite small in the IA model, thus the diffusion is dominated by the balance of the pressure diffusion force and the ordinary diffusion force. In Fig 5.14(c), a good match can be observed when comparing the thermal segregation profiles between the JM model and IA model. From a zero value at the center the thermal segregation increases quickly, passes through a maximum and decreases quickly until its value is close to zero, then reaches a plateau near the walls. The positive values of the thermal segregation near the center indicates that small particles have a more capability to migrate towards the center than large particles, leading to a more non-uniform distribution of small particles.

By comparing the results of the equal mass case and the equal density case, we can see that the solids species segregation has a transition. For the equal density mixture, large particles have more potential to segregate in high density regions. For the equal mass mixture, the situation is reversed. To better understand this transition, the variation of solids fractions and corresponding diffusion forces with the mass ratio is plotted in Fig 5.15 when the size ratio is held constant at 2. It is clear that a segregation reversal is observed with an increase in the mass ratio. For small mass ratios, small particles segregate to the high density regions. For large mass ratios, large particles segregate to the high density regions. It is important to note that corresponding to the reversal of segregation the pressure diffusion also has a reversal. For small mass ratios, the pressure diffusion force is positive and the thermal diffusion force is quite small from $y^*=0$ to 1, driving small particles to segregate to the high density regions. With an increase in the mass ratio the direction of the pressure diffusion force reverses and the contribution of the thermal diffusion force becomes pronounced, leading to large particles to segregate in the high density regions. Similar results were obtained by Khakhar et al. (1999) for flow down chutes.

In Fig 5.16(a) the variation of the average granular energy with the size ratio is plotted for the JM model and the IA model respectively. As for the equal density case (see Fig 5.8(a)), the average granular energy decreases with an increase in the size ratio. However, the decrease in the equal mass case is much slower than the equal density case. The variation of the granular energy ratio with R is shown in Fig 5.16 (b). Although the granular energy is unequally partitioned between two species, the degree of non-equipartition is rather small comparing with the equal density case (see Fig 5.8 (b)). In addition, over the range of the size ratios studied the granular energy of large particles is smaller than small particles for the simple shear flow, which is the opposite of what was observed in the equal density case. Therefore, we cannot use the size disparity to explain the non-equipartition of granular energy obtained in the equal density case. A further study of the effect of the mass disparity is required. It is of interest to note that the average granular energy ratio for the non-simple shear flow exhibits a non-monotonic behavior with the variation of the size ratio. This is related to the solids fraction and granular energy distributions. For small size ratios, the segregation between two particle species is not significant thus the energy ratio is mainly dominated by the granular energy distribution. From previous results, we have known that for the equal mass mixture the granular energy of small particles is higher than that of large particles so that the energy ratio is less than 1. With an increase in the size ratio, the species segregation becomes significant, and small particles accumulate in dense regions with a low granular energy and large particles distribute to dilute regions with a high granular energy. Therefore, the energy ratio increases with an increase in the size ratio. This non-monotonic behavior is not shown in Fig 5.5(b) and Fig 5.8(b) for the equal density mixtures. This is because under the dilute situations studied in the equal density mixtures the contribution of the granular energy distribution is more pronounced than the solids species segregation so that the energy ratio is larger than 1. For high mean solids fractions, the species segregation is significant and the average granular energy of large particles may be lower than that of small particles for the equal density particles.

5.2.1.3 Equal Size Mixture

In this section, we set the particle size equal to each other (d_A=d_B), and investigate the effect of the mass ratio (M) on granular flow behavior and segregation. The flow profiles of solids fraction, granular energy and the solids velocity with different mass ratios are shown in Fig 5.17, for the JM model. Although the total solids segregation increases with an increase in the mass ratio, the effect of the mass ratio on the solids species segregation is more significant. For M=2, both heavy and light particles monotonically decrease their concentrations from the center to the walls. When the mass ratio increases to 8, the concentration of light particles increases from the center, passes through a maximum and then decreases towards the walls, while the concentration of large particles monotonically decreases from the center to the walls. Therefore, a significant solids species segregation can be found near the center and this segregation is further enhanced when the mass ratio increases to 64. The granular energy decreases with an increase in the mass ratio. The effect of the mass ratio on the solids velocity profile is insignificant.

Similar profiles are observed for the IA model when the mass ratio increases from 2 to 8 (see Fig 5.18(a)). However, when the mass ratio further increases from 8 to 64, the situation reverses in the large particle phase. With increasing the mass ratio the solids fraction of large particles decreases in the region near the center, leading to a decrease in the total solids segregation as well as the solids species segregation. This is not observed in the JM model. The profiles of the granular energies are shown in Fig 5.18(c) and (d). Although the granular energy of small particles decreases with an increase in the mass ratio, the variation of the granular energy of large particles is non-monotonic. Initially the granular energy of large particles decreases with increasing the mass ratio, after a threshold (M=16 under our specific situation) it starts to increase. This is in agreement with the solids fraction profile of large particles. The effect of the mass ratio on the solids velocity profiles is not significant.

Fig 5.19 shows the profiles of the diffusion forces with different mass ratios based on the JM model and the IA model respectively. For the JM model, the contribution of the thermal diffusion force and the ordinary diffusion force is compensated by the pressure diffusion force, and most of changes in the diffusion forces can be observed near the center. These profiles are quite similar with those obtained for the equal density case (see Fig 5.12(b)). With an increase in the mass ratio, the absolute values of the three diffusion forces increase and the contribution of the ordinary diffusion force becomes more important. For the IA model, when the mass ratio increases from 2 to 8 the absolute value of three diffusion forces increase and the growth of the ordinary diffusion force is faster than the thermal diffusion force. With continuously increasing the mass ratio from 8 to 64, the absolute values of the three diffusion forces decrease and the decrease in the ordinary diffusion force is faster than the thermal diffusion force. This is in agreement with the non-monotonic flow profiles observed in Fig 5.18 (IA model).

To further investigate the physical mechanism leading to the segregation in the equal size case, the profiles of the thermal diffusion factors and the thermal segregation are plotted in Fig 5.20. For the JM model, the thermal diffusion factor is positive across the gap and increases with an increase in the mass ratio. Thus heavy particles segregate to the low granular energy regions and the solids species segregation increases with an increase in the mass ratio. For the IA model, although the thermal factor profile is qualitatively similar to that in the equal density case (see Fig 5.11(a) for the IA model), it is interesting to note that the thermal segregation has a transition with the variation of the mass ratio. When the mass ratio is small the absolute values of the thermal factors increase and the magnitude of the thermal segregation decreases with an increase in the mass ratio (not shown in the figures), indicating that the solids species segregation increases with an increase in the mass ratio. When the mass ratio is above a threshold, the situation reverses. From Fig 5.20(c) and (d), it can be seen that when the mass ratio increases from 8 to 64 the absolute values of the thermal factors decrease and the magnitude of the thermal segregation increases near the center, resulting in a decrease in the solids species segregation in the system.

The variation of the average granular energy with the mass ratio is plotted in Fig 5.21(a). For the JM model, the average granular energy decreases with an increase in the mass ratio. A similar monotonic behavior is observed for the IA model when plotting the variation of the average granular energy of small particles versus the mass ratio. For large particles in the IA model, however, the granular energy exhibits a non-monotonic behavior with the mass ratio. As the mass ratio increases from 1, the granular energy decreases quickly, passes through a minimum, and then increases gradually. This non-monotonic profile was also found in particle dynamic simulations (Alam and Luding, 2002a). To find out how the granular energy is partitioned between the two species in the equal size case, the variation of the granular energy ratio with the mass ratio is shown in Fig 5.21 (b). It can be seen that the granular energy ratio linearly increases with increasing the mass ratio, and the slope for the non-simple shear flow is larger than the simple shear flow. By comparing Fig 5.16(b) and Fig 5.21(b), it is clear that the effect of the mass disparity on the breakdown of equipartition of granular energy is much more significant than the size disparity. Therefore, we can conclude that the breakdown of equipartition of granular energy in the equal density case (see Fig 5.8(b)) is mainly due to the contribution of the mass disparity.

5.2.2 Granular Channel Flow

An understanding of granular material flowing under gravity is essential to the solution of many problems associated with the handling and movement of these materials. Previous work on flow between vertical parallel plates has shown that the assembly of particles behaves quite differently from a Newtonian fluid, manifesting dense regions in the center and dilute regions near the walls (Wang et al., 1997). In this section, the effect of physical parameters and operating conditions on the flow behavior and segregation is investigated for granular flow in a channel with bi-disperse particles.

5.2.2.1 Effect of the restitution coefficient for the equal density case

The contribution of particle inelasticity on granular flows and gas-particle flows has been discussed in a number of papers (Goldhirsch and Zanetti, 1993; Hrenya and Sinclair, 1997; Liu and Glasser, 2006; Tan and Goldhirsch, 1997). To compare and contrast the rheology of a bidisperse granular medium with that of its monodisperse counterpart, we first present our results for the mono-disperse case. The solids fraction, granular energy and solids velocity profiles for different coefficient of particle-particle restitution for monodisperse particles are shown in Fig 5.22, using a kinetic theory model based on the work of Iddir and Arastoopour (2005). As discussed in 5.1.2, the velocity and granular energy have been non-dimensionlized by $\sqrt{g\Delta}$ and $\frac{1}{3}m_A(\sqrt{g\Delta})^2$. The ratio of gap width to the large particle diameter is 40, and the particle-wall restitution coefficient is 0.9.

When elastic collisions between particles are assumed (e_p=1), there is a plateau of low solids concentration in the center and a relatively high concentration of particles near the walls. The granular energy has a plateau in the center region and decreases close to the walls due to the energy loss at the walls (e_w=0.9). When the granular system changes from an elastic case to a slightly inelastic one, the particles start to migrate from the walls to the center. For e_p=0.96, there is a high concentration of particles near the center and a dilute region near the walls. The granular energy increases when moving away from the center, passes through a maximum and then decreases near the walls. This sensitivity of flow profiles to the coefficient of particle-particle restitution has been discussed in previous works (Hrenya and Sinclair, 1997; Liu and Glasser, 2006; Louge et al., 1991; Ocone et al., 1993; Pita and Sundaresan, 1991; Sinclair and Jackson, 1989). The solids velocity profiles are shown in Fig 5.22(c). The velocity is higher at the center and lower at the walls, and its magnitude increases with a decrease in the restitution coefficient. We believe this is related to the solids fraction distribution. In the center region the solids fraction increases with a decrease in e_p, leading to a higher system density in the center. As for fluids, the velocity at the center increases with an increase in density (Bird et al., 1960).

The granular flow profiles for an equal density particle mixture are shown in Fig 5.23, when elastic particle-particle collisions are assumed. Generally, the flow profiles are similar to the monodisperse case. Both large and small particle species have a low concentration in the center region and a high concentration at the walls (see Fig 5.23(a)). However, the distribution of two particle species is not equal. The inhomogeneity in the large particle phase is much more pronounced than that in the small particle phase. The granular energies of both large and small particles are nearly identical and have a large magnitude at the center and decrease towards the walls (see Fig 5.23(b)). This is in an agreement with the analysis of Barrat and Trizac (2002), who showed for elastic particles (e_p=1) there is no difference between the granular energy of large and small particles. The solids velocity profiles are shown in Fig 5.23(c). Clearly, across the channel the two particle species have almost identical velocities. This is because F^*_{D1i} in Equation 5.13 can be rewritten as $C^*_{FD1i}(v^*_j - v^*_i)$ and a large value of C^*_{FD1i} is obtained in the calculation. All other terms in Equation 5.13 are relatively small. Therefore, the relative velocity between two particle species has to be small.

From Fig 5.23(a), it is clear that there is a segregation and large particles accumulate preferentially near the walls. To further investigate the physical mechanism leading to this segregation, the diffusion equation is investigated based on the work of Iddir and Arastoopour (2005). The expressions for the diffusion coefficient and the diffusion forces are given by Equation 5.28 and 5.29. From Fig 5.23(c), we have shown that the solids velocities of two particle species are almost equal (This is also true for other cases we carried out. See the results in the following section for detail). The definition of the diffusion velocity of particle species i is given by $\underline{v}_i' = \underline{v}_i - \bar{\underline{v}}$, where $\bar{\underline{v}} = \frac{\phi_A v_A + \phi_B v_B}{\phi_A + \phi_B}$. Since $\underline{v}_A \approx \underline{v}_B$, $\bar{\underline{v}} \approx \underline{v}_A$. Therefore, it is reasonable to assume that $\underline{v}_A' - \underline{v}_B' = \underline{0}$ leading to $\underline{d}_1 = \underline{0}$. Thus, the segregation between two particle species is mainly due to the competition of the pressure diffusion force, the thermal diffusion force and the ordinary diffusion force.

The ordinary diffusion coefficient profile and the diffusion force profiles are shown in Fig 5.24, corresponding to the flow profiles shown in Fig 5.23. Since D_{AB}^* decreases with an increase in the solids fraction and increases with an increase in the granular energy, a monotonous decrease in D_{AB}^* is predicted from the center to the walls corresponding to the profiles of the solids fraction and the granular energy. The profiles of three diffusion forces have plateaus near the center, and exhibit significant changes close to the walls, where the combination of the ordinary diffusion force and the thermal diffusion force is compensated by the pressure diffusion force. Similar results were obtained for gravity-free particle dynamic simulations (Galvin et al., 2005) between a hot wall and a cold wall. Since the ordinary diffusion force is proportional to $-(\frac{\nabla n_A}{n_A} - \frac{\nabla n_B}{n_B})$ (see Equation 5.29), a big change of the ordinary diffusion force observed near the walls leads to a significant segregation between two particle species in that region. In addition, the sign of the ordinary diffusion force determines the direction of the solids migration. At y*=1, a negative ordinary diffusion force near the walls indicates that large particles have a higher propensity to migrate towards the walls so that the solids fraction of large particles is much higher than that of small particles close to the walls.

When the particle-particle restitution coefficient decreases to 0.99, the solids fraction distribution reverses. Both large and small particles have a high concentration in the center region and a low concentration near the walls (see Fig 5.25(a)). The change in the solids fraction profile of large particles is much more significant than that of the small particles. The granular energy montonically increases from the center to the walls. Since the collisions between particles are now inelastic, the equi-partition of granular energy breaks down, and the granular energy of large particles is slightly higher than small particles across the channel (see Fig 5.25(b)). This behavior has also been captured using particle dynamic simulations for Couette flows (Alam and Luding, 2002a, b, 2003). The solids velocity profiles are shown in Fig 5.25(c), in which both particle species have almost identical velocities across the channel. Scrutiny of Fig 5.23(c) and Fig 5.25(c) reveals that for the bidisperse particle system the solids velocities increase with a decrease in the particle-particle restitution coefficient, which is in agreement with the results for the monodisperse case.

In Fig 5.25(a), it is clear that a segregation occurs between two particle species. To better understand the physical mechanism leading to this segregation, the diffusion profiles are plotted in Fig 5.26. The ordinary diffusion coefficient has a minimum at the center and montonically increases towards the walls. The diffusion force profiles exhibit plateaus near the walls, and most of changes can be observed near the center, corresponding to the flow profiles in Fig 5.25. It is important to note that corresponding to the transition of the solids fraction distribution with the variation of e_p, the diffusion force profiles also exhibit a transition. When e_p=1, the pressure diffusion force has positive values from y*=0 to 1 and the thermal diffusion force and the ordinary diffusion force are negative, causing large particles to migrate towards the walls. When e_p decreases to 0.99, this situation reverses, driving large particles to segregate to the center.

To further investigate the transition phenomenon observed in the previous results, the profiles of an intermediate state is plotted in Fig 5.27, where the coefficient of particle-particle restitution is equal to 0.997. By comparing Fig 5.23(a), Fig 5.25(a) and Fig 5.27(a), it is clear that a transition occurs in the solids fraction distribution, and the transition in the large particle phase is much more significant than the small particle phase. A transition is also observed in the granular energy profiles (see Fig 5.23(b), Fig 5.25(b) and Fig 5.27(b)). When e_p=1, the pseudo-thermal energy has a flux from the center to the walls. With decreasing e_p, the situation reverses, and the pseudo-thermal energy starts to flow from the walls to the center. The diffusion force profiles for the intermediate state are shown in Fig 5.27(c). The distribution of three diffusion forces is similar to the elastic case (e_p=1) in the region close to the walls. When moving towards the center, the values of the three diffusion forces pass through zero respectively and the profiles exhibit a similarity to the case of e_p=0.99.

When e_p further decreases, the sensitivity of the flow profiles to this parameter is reduced (see Fig 5.28), which is in agreement with the simulation results obtained in the monodisperse systems. The segregation between two particle species is enhanced with a decrease in the restitution coefficient. In Fig 5.28(a), the solids fraction of large particles has a maximum at the center and montonically decreases towards the walls, while the solids fraction of small particles increases from the center, passes through a maximum and then decreases towards the walls, leading to a segregation phenomenon observed in the center region. This is due to the big jump of the three diffusion forces near the center, where a large magnitude of the ordinary diffusion force leads to a high concentration of large particles at the center. (The diffusion force profiles are similar to Fig 5.26(b) except that the magnitudes are much larger near the center, and thus are not shown here.) The breakdown of equipartition of granular energy is also enhanced with decreasing e_p. (see Fig 5.28(b)) This behavior has been observed in particle dynamic simulations (Alam and Luding, 2002b). The solids velocity profiles of the two particle species are quite similar except that a small difference can be observed near the walls.

From Fig 5.22, we have seen that the flow profiles in the monodisperse case are sensitive to the coefficient of particle-particle restitution when its value is close to 1. The effect of this parameter on the total solids fraction profile in the bidesperse particle system is shown in Fig 5.29. By comparing Fig 5.22(a) and Fig 5.29(a), it can be seen that under this condition the solids fraction profiles are much more sensitive to the variation of the restitution coefficient in the bidisperse case than that in the monodisperse case. This is related to the particle species used in the simulation. In the monodisperse case, particle species A is used with average solids fraction equal to 0.1. In the bidisperse case, particle species A and B are used with average solids fraction equal to 0.05 respectively. Since $d_B=0.5d_A$ and the physical properties of particle species A are used for the dimensionless calculation, the average number density in the bidisperse case is 4.5

times that of the monodisperse case, leading to the frequency of particle-particle collisions in the bidisperse case to be much higher than the monodisperse case. Therefore, the flow profiles are more sensitive to the particle-particle restitution coefficient in the bidisperse case. Based on the above analysis, it is expected that in a bidisperse mixture if particle species B is replaced with another particle species C with $d_C=1.5d_A$ and $\phi_A=\phi_c=0.05$, the sensitivity to e_p should be reduced. Fig 5.29(b) confirms that when particles A and C are mixed, the sensitivity of the total solids fraction to e_p is greatly suppressed.

5.2.2.2 Effect of the mean solids fraction for the equal density case

From previous work, it is known that for a monodisperse granular flow the average solids load has a significant effect on the flow profiles and instabilities (Stephen, 2006; Wang et al., 1997). It is of interest to extend the work to a bidisperse system to investigate its effect on the flow profiles and segregation behavior. Solutions for an increased total average solids fraction have been computed. The solids fraction profiles and corresponding diffusion force profiles are shown in Fig 5.30. For the elastic system (see Fig 5.30(a)), the concentration profiles of both large particles and small particles exhibit a plateau near the center. When moving towards the walls, the solids fraction of large particles increase while the solids fraction of small particles montonically decreases, leading to a pronounced demixing near the walls that is not observed in the dilute system ($\bar{\phi}<0.1$). When particle-particle collisions are inelastic (see Fig 5.30(b)), a significant segregation between two particle species can be observed near the center. By comparing with Fig 5.25(a), we can see that the segregation is greatly enhanced with an increase in the solids load. In the dilute system, small particles have a minimum concentration at the walls, while the situation reverses in the dense system where the minimum concentration of small particles is found at the center. By comparing Fig 5.24(b) and Fig 5.30(c), Fig 5.26(b) and Fig 5.30(d), we find that the diffusion force profiles look similar between the dilute system and the dense system except that the magnitude (absolute value) of the three diffusion forces has a significant increase with an increase in the system solids load, resulting in a much more pronounced segregation observed in the dense system.

5.2.2.3 Effect of the size ratio for the equal density case

The next parameter considered is the size ratio. Fig 5.31(a) and (b) show that the solids distributions are sensitive to the variation of the size ratio. When the size ratio increases from 2 to 3, the total solids fraction increases significantly at the center and a much more dilute region can be observed close to the walls. The segregation between the two particle species is also enhanced with an increase in this parameter. It is interesting to note that under this relatively low solids load ($\bar{\phi}_A=\bar{\phi}_B=0.025$), there is no segregation observed between the two particle species when the size ratio is equal to 2. Recalling the results shown in Fig 5.25(a) (R=2, $\bar{\phi}$=0.1) and Fig 5.30(b) (R=2, $\bar{\phi}$=0.2), it is clear that the solids species segregation is greatly enhanced with an increase in the solids load in the system. The profiles of the granular energies are shown in Fig 5.31(c) and (d). With an increase in the size ratio, the granular energies of both large and small particles decrease in the region near the center and increase in the region close to the walls. This corresponds to the distribution of the solids fraction because the higher the solids fraction the lower the granular energy. By comparing Fig 5.31(c) and (d), we can see that the breakdown of equipartition of granular energy is enhanced with an increase in the size ratio. This result has been observed in particle dynamic simulations (Alam and Luding, 2002b). For the equal density cases, we find that under our specific conditions the granular energy of large particles is always higher than the small particles when the collisions between particles are inelastic. In addition, although the particle inelasticity alone can lead to the non-equipartition of granular energy, the effect of the size disparity is more significant. Similar results have been observed in Couette granular flows. The solids velocity profiles for different size ratios are shown in Fig 5.31(e) and (f). Clearly, the solids velocities of both particle species increase with an increase in the size ratio.

5.2.2.4 Effect of the size ratio for the equal mass case

So far our study has been restricted to the equal density case where the two particle species are made from the same material but differ in size. Thus, a difference in the size ratio automatically leads to a difference in the mass ratio. In granular Couette flow, we know that the mass disparity and size disparity have different effects on the granular flow. To study whether these features can be also captured in granular channel flows, a further investigation of the equal mass case and the equal size case in the channel flow is required.

In Fig 5.32, we set the particle masses equal to each other ($m_A=m_B$), and investigate the effect of the size ratio (R) on the granular flow. It is found that the segregation between two particle species is quite different from the equal density mixture. When two particle species have the same density, the solids distribution of large particles is more heterogeneous than that of small particles, especially in the region near the center where the solids fraction of large particles can be much higher than that of small particles (see Fig 5.31). In the equal mass mixture, the situation reverses (see Figure 5.32(a) and (b)). At the center, small particles have a higher solids fraction than large particles. Near the walls, the concentration of small particles increases slightly due to the sink wall condition, while the concentration of large particles decreases. With an increase in the size ratio, the solids fraction of large particles decreases while the solids fraction of small particles increases in the center, leading to a more pronounced species segregation observed in that region. The granular energy profiles are shown in Fig 5.32(c) and (d). It is important to note that the granular energy of the large and small particles is almost equal and this feature does not change with the variation of the size ratio. Therefore, the assumption of equi-partition of granular energy can be applied under the equal mass situation. Similar profiles of solids fraction and granular energy have been observed in Couette granular flows. With an increase in the size ratio the granular energy increases across the channel except in the region near the center where the situation reverses. By comparing with the granular energy profiles for the equal density case, it is found that the granular energies in the equal mass case have a more potential to increase with an increase in the size ratio. This is because the value of the granular energy is proportional to the particle mass. In the equal density case the mass of small particles decreases with an increase in the size ratio, while in the equal mass case the mass of small particles does not change with the variation of the size ratio leading to an increase in the system energy. The solids velocities at the center increase with an increase in the size ratio due to a higher solids fraction observed in that region.

The diffusion profiles are shown in Fig 5.33, corresponding to the flow profiles observed in Fig 5.32. The ordinary diffusion coefficient decreases with an increase in the size ratio because this coefficient is proportional to $\frac{R^2}{(1+R^2)(\phi_A+\phi_B R^3)}$ in the equal mass case. The pressure diffusion force increases from the center, passes a maximum and then decreases. The profile of the ordinary diffusion force is reversed. By comparing Fig 5.33 with Fig 5.24 and Fig 5.26, we can see that the directions of the pressure diffusion forces and the ordinary diffusion forces are opposite between the equal density case and the equal mass case. Although the thermal diffusion force has a positive value across the channel in both the equal density and equal mass mixtures, its magnitude is much smaller in the equal mass case than that in the equal density case due to the similarity of the granular energy profiles between the two particle species in the equal mass case.

By comparing the results of the equal mass case and equal density case, we can see that the segregation between two particle species has a reversal. For the equal density case, large particles have more potential to segregate in dense regions. For the equal mass case, the situation reverses. Similar results are obtained in Couette granular flows.

5.2.2.5 Effect of the mass ratio for the equal size case

The effect of the mass ratio (M) on the granular flow behavior is shown in Fig 5.34 when the particle sizes are assumed to be equal to each other (d_A=d_B). The effect of the mass ratio on the solids fraction profile is significant. For M=2, both heavy and light particles montonically decrease their concentrations from the center to the walls. When the mass ratio increases to 8, the concentration of light particles increases from the center, passes through a maximum and then decreases towards the walls, while the concentration of large particles montonically decreases from the center to the walls. Therefore, a demixing phenomenon can be found near the center and this segregation is further enhanced when the mass ratio increases to 32. It is interesting to note that when the mass ratio further increases from 32 to 64, the concentration of large particles start to decrease in the center, leading to a decrease in the segregation observed in the solids phase. This non-monotonic profile was also found in particle dynamic simulations (Alam and Luding, 2002a) and the results of granular Couette flows in the previous section. By comparing the solids fraction profiles in Fig 5.31, 5.32 and 5.34, we find a competition mechanism exists in the equal density case since in the equal mass case small particles have a higher solids fraction in the dense regions while in the equal size case the solids fraction of heavy particles is higher in the dense regions. Under our specific situation large particles have more potential to segregate in the dense regions in the equal density case indicating that the contribution of the mass disparity is more significant than the effect of the size disparity. Similar behavior has been found in granular Couette flows. The granular energies of both large and small particles (see Fig 5.34(c) and (d)) decrease with an increase in the mass ratio across the channel except in the region near the walls where the situation is reversed for the large particles. This is related to the solids fraction distribution and the definition of the granular energy because the granular energy decreases with an increase in the solids fraction and in this equal size case the system energy decreases with an increase in the mass ratio. The breakdown of equipartition of granular energy is enhanced with an increase in the mass ratio. Scrutiny of the granular energy profiles in Fig 5.31, 5.32 and 5.34 reveals that the distribution of the granular energy in the equal density case is attributed to the competition of the size disparity and the mass disparity, and the non-equipartition of granular energy observed in the equal density case is mainly due to the effect of the mass disparity. Both solids velocities increase when the mass ratio increases from 2 to 32 (see Fig 5.34(e) and (f)). With a further increase in the mass ratio, however, its effect on the solids velocity profile becomes small.

The diffusion profiles for the equal size case are shown in Fig 5.35, corresponding to the flow profiles observed in Fig 5.34. The ordinary diffusion coefficient increases with an increase in the mass ratio because this coefficient is proportional to $(1+M)^{0.5}$ in the equal size case if the level of non-equipartition of granular energy is not pronounced. The profiles of the pressure diffusion force and the ordinary diffusion force are qualitatively similar to the equal density case, and the magnitude of the thermal diffusion force is always positive across the channel. It is important to note that corresponding to the transition observed in the segregation between two particle species a transition also occurs in the diffusion force profiles with the variation of the mass ratio. When the mass ratio increases from 2 to 32, the absolute values of the three diffusion forces increases. As the mass ratio further increases, this situation reverses. Therefore, it is clear that the non-monotonic behavior observed in the flow profiles can be attributed to the non-monotonic change of the diffusion force magnitudes.

5.2.3 Gas-Particle Flow in a Channel

Finally, we extend our work to a gas-particle channel system with a bimodal particle mixture to investigate how the gas affects the flow hydrodynamics. It is worth mentioning that in the granular system the positive vertical direction is downward while in the gas-particle system the positive vertical direction is upward.

When the collisions between particles are elastic, the solids fraction and granular energy profiles of gas-particle flows (see Fig 5.36(a) and (b)) are quite similar to the granular flows (see Fig 5.23), although the magnitudes of the granular energies are higher in the granular system than the gas-particle system. This is due to the similarity of the diffusion force profiles between the two systems (see Fig 5.24(b) and Fig 5.36(d)). The velocity profiles are shown in Fig 5.36(c). Both the gas and solids velocities have a maximum at the center and montonically decrease from the center to the walls. Similar profiles are obtained in the gas-particle flow with a monodisperse particle mixture (Liu and Glasser, 2006). Scrutiny of Fig 5.23(c) and Fig 5.36(c) indicates that the dimensionless solids velocities are higher in the granular flows than the gas-particle flows, which leads to the higher granular energies observed in the granular system than the gas-particle system.

The profiles of the gas-particle flows and their corresponding diffusion force profiles are plotted in Fig 5.37 when the coefficient of the particle-particle restitution is 0.99. By comparing Fig 5.37 with Fig 5.25 and Fig 5.26(b), it can be seen that under this inelastic situation the similarity of the solids fraction and the granular energy distribution and the diffusion force profiles also exists between the gas-particle system and the granular system. Thus, it is clear that for the kinetic theory models used in the present work the effect of the gas is mainly in the vertical direction, driving the particles to move upward, and its effect on the solids distribution across the channel is not significant because there is no gas related terms in the solids lateral momentum equations. Scrutiny of Fig 5.25(c) and 5.37(c) reveals that the solids velocities are higher in the granular flows than the gas-particle flows, leading to the higher granular energies observed in the granular system. This is in agreement with the results obtained in the elastic system. It is interesting to note that if collisions between particles are inelastic a big jump occurs in the gas velocity profile leading to a large magnitude of the relative velocity between gas and solids in the region close to the walls. Similar flow profiles were obtained in gas-particle systems with a monodisperse particle mixture (Liu and Glasser, 2006; Sinclair, 1989).

5.3 Conclusions

In this chapter, the hydrodynamics of granular flows and gas-particle flows with a bi-disperse particle mixture are investigated based on kinetic theory. To investigate the effect of non-equipartition of granular energy, two continuum models are investigated. One is based on the work of Jenkins and Mancini (1989), and the other is based on the work of Iddir and Arastoopour (2005). The steady state solutions are obtained using the numerical method of orthogonal collocation.

For an equal density case, the breakdown of equipartition of granular energy is enhanced with a decrease in the system inelasticity, an increase in the average solids fraction or an increase in the size ratio between two particle species. By investigating an equal size and equal mass mixture, we find that the breakdown of energy equipartition is mainly due to the contribution of the mass disparity, whereas the effect of the size disparity is very small if the two particle species have the same mass. Therefore, for the equal mass case, the granular flows exhibit similar profiles between the JM model and IA model.

In granular systems, the segregation of total particles and each individual particle species is enhanced when the effect of non-equipartition of granular energy is considered. The species segregation in the solids phase is due to the competition of three diffusion forces: the thermal diffusion force, the ordinary diffusion force and the pressure diffusion force. In addition, we find a competition mechanism exists in the equal density case since in the equal mass case small particles have a higher concentration in low energy regions whereas in the equal size case the solids fraction of heavy particles is higher in low energy regions.

In granular channel flows, the flow profiles are sensitive to the coefficient of particle-particle restitution when its value is close to 1. The concentration of both particle species has a transition from accumulation close to the walls to accumulation near the center when e_p decreases from 1 to 0.96. This sensitivity is reduced when the system becomes more inelastic. By comparing with the monodisperse system, we find that if larger particles are mixed in the system the sensitivity to the restitution coefficient is suppressed, while if smaller particles are added in the system the situation reverses.

Although granular Couette flow and granular channel flow exhibit similar profiles for solids fraction and granular energy, the solids velocity profiles are quite different. For granular Couette flow, the range of the dimensionless solids velocities is from zero at the center to unity at the walls. This constraint determines that the solids velocity profiles are not sensitive to the variation of the parameters. For granular channel flow, the solids velocity is not limited by the boundary conditions. Therefore, the parametric effects on the solids velocity profiles become significant. The solids velocity increases with an increase in the system inelasticity, the average solids load, the size ratio and the mass ratio.

Then we extend our work to gas-particle flows in a channel. By comparing with the results in granular channel flows, we find that for the kinetic theory models used in the present study, the solids fraction and granular energy profiles are quite similar between the granular flows and the gas-particle flows, indicating that the effect of the gas is mainly in the vertical direction, which drives particles to be suspended in the bed.

Fig 5.1: Schematic diagram of a bounded granular Couette flow.

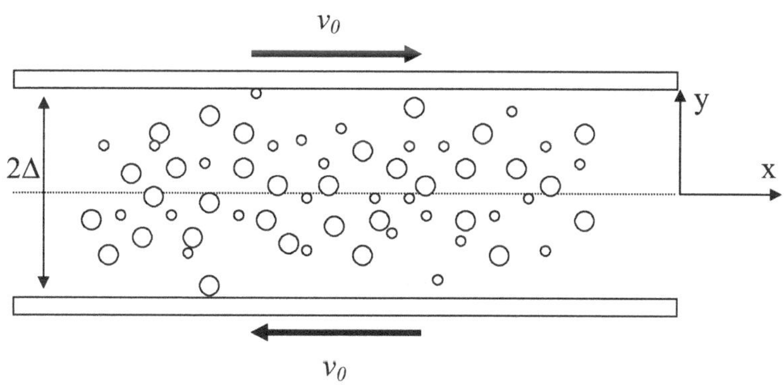

Fig 5.2: Schematic diagram of a bounded granular channel flow.

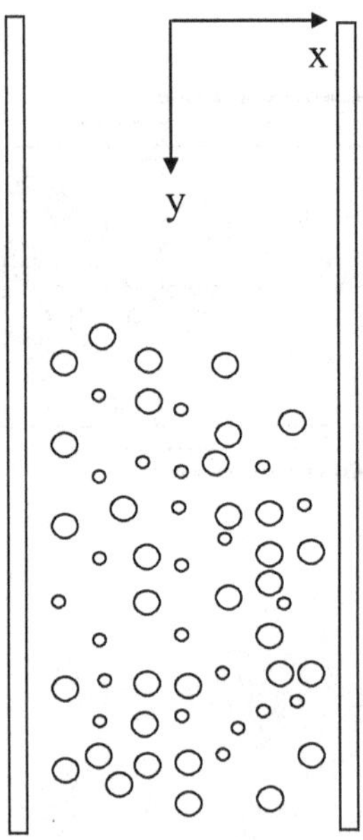

Fig 5.3: Effect of the coefficient of restitution on the flow profiles for the JM model ($\bar{\phi}$=0.02, χ=1, R=2, $\rho_A=\rho_B$). a) solids fraction profile of large particles. b) solids fraction profile of small particles. c) granular energy profile. d) velocity profile.

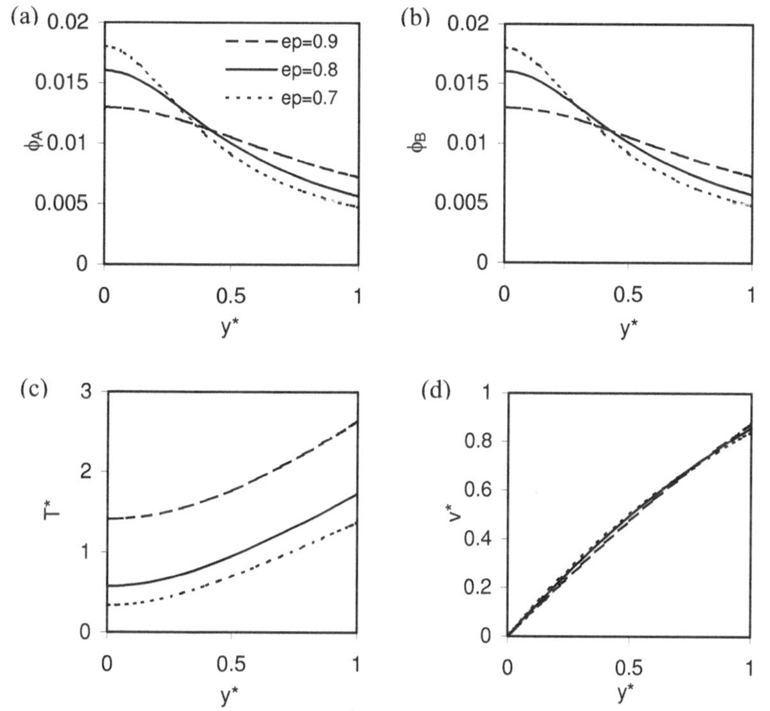

Fig 5.4: Effect of the coefficient of restitution on the flow profiles for the IA model ($\bar{\phi}$=0.02, χ=1, R=2, ρ_A=ρ_B). a) solids fraction profile of large particles. b) solids fraction profile of small particles. c) granular energy profile of large particles. d) granular energy profile of small particles. e) velocity profile.

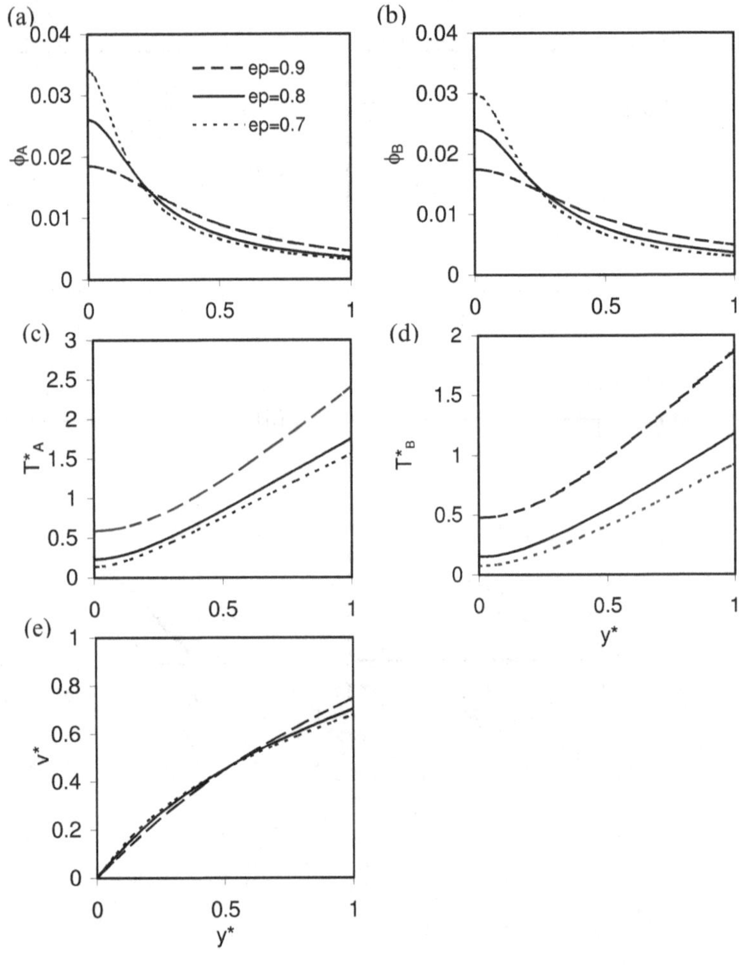

Fig 5.5: Effect of the coefficient of restitution on the average granular energy ($\bar{\phi}$=0.02, χ=1, R=2, ρ_A=ρ_B). a) variation of average granular energy with the coefficient of restitution. b) variation of granular energy ratio with the coefficient of restitution.

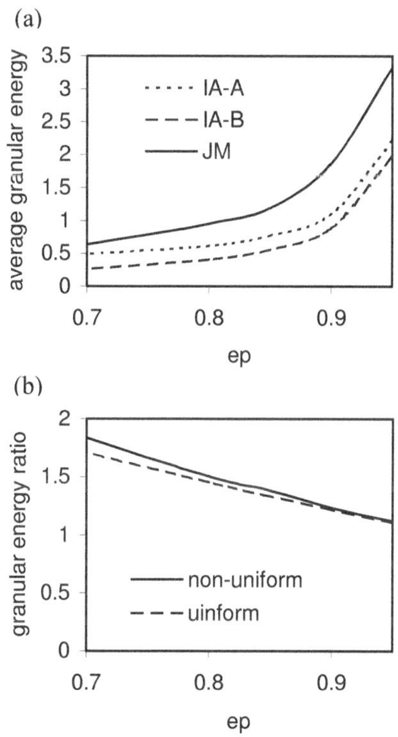

Fig 5.6: Effect of the size ratio on the flow profiles for the JM model ($\bar{\phi}$=0.02, χ=1, e_p=0.9, ρ_A=ρ_B). a) solids fraction profile of large particles. b) solids fraction profile of small particles. c) granular energy profile. d) velocity profile.

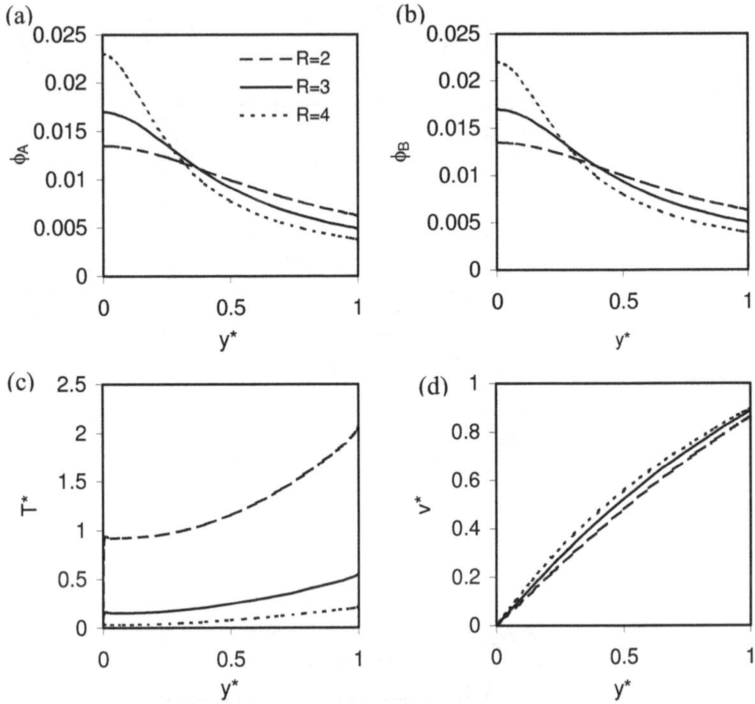

Fig 5.7: Effect of the size ratio on the flow profiles for the IA model ($\overline{\phi}$=0.02, χ=1, e_p=0.9, ρ_A=ρ_B). a) solids fraction profile of large particles. b) solids fraction profile of small particles. c) granular energy profile of large particles. d) granular energy profile of small particles. e) velocity profile.

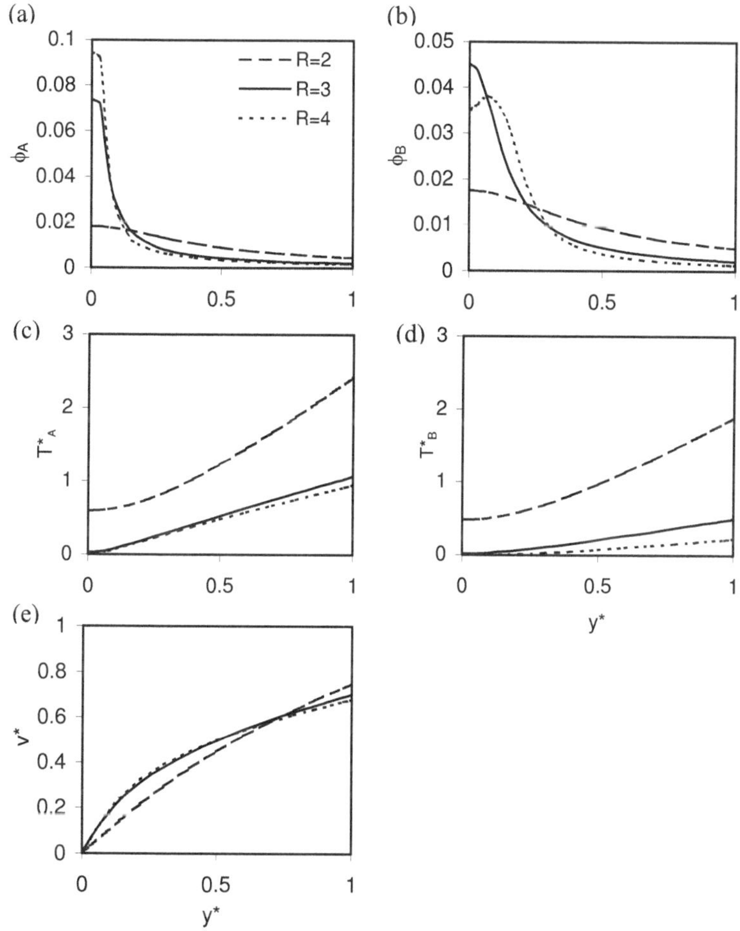

Fig 5.8: Effect of the size ratio on the average granular energy ($\bar{\phi}$=0.02, χ=1, e_p=0.9, ρ_A=ρ_B). a) variation of average granular energy with the size ratio. b) variation of granular energy ratio with the size ratio.

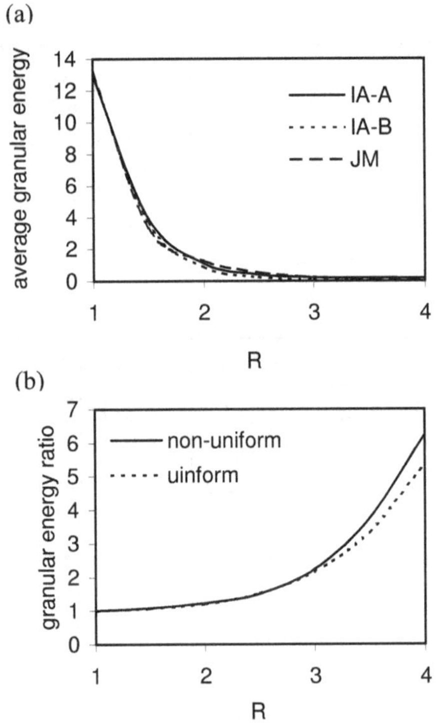

Fig 5.9: Comparison of flow profiles between the JM model and IA model ($\bar{\phi}$=0.04, χ=1, e_p=0.9, R=2, $\rho_A=\rho_B$). a) solids fraction profile. b) granular energy profile. c) velocity profile.

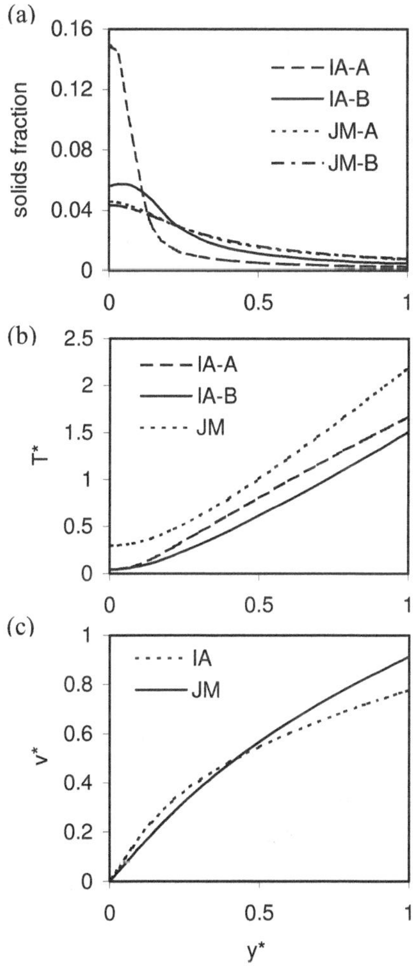

Fig 5.10: Comparison of the diffusion profiles between the JM model and IA model ($\bar{\phi}$=0.04, χ=1, e_p=0.9, R=2, ρ_A=ρ_B). a) ordinary diffusion coefficient profile. b) diffusion force profile, for the JM model. c) diffusion force profile, for the IA model.

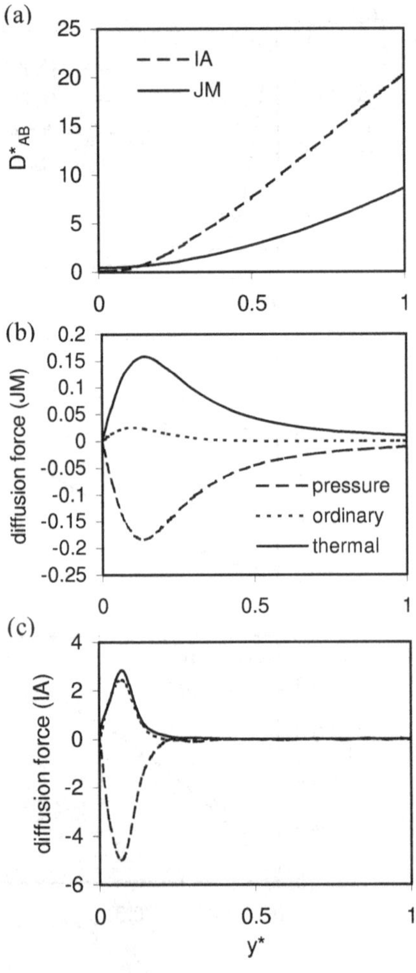

Fig 5.11: Comparison of the thermal segregation profiles between the JM model and IA model ($\bar{\phi}$=0.04, χ=1, e_p=0.9, R=2, ρ_A=ρ_B). a) thermal diffusion factor profile. b) thermal segregation profile.

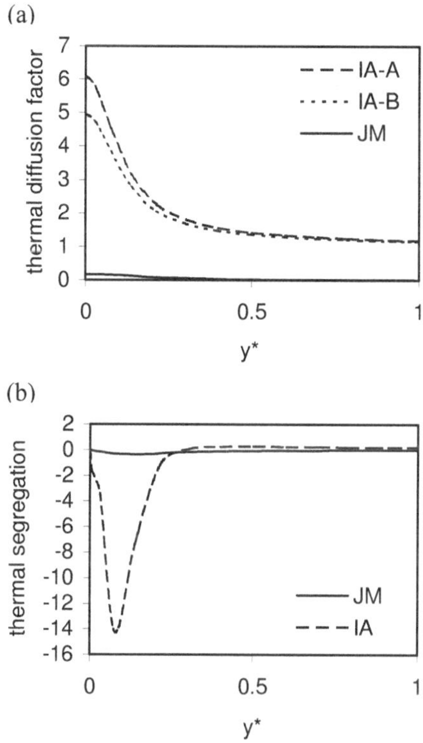

Fig 5.12: Flow and diffusion profiles with a higher solids fraction, for the JM model ($\bar{\phi}$=0.06, χ=1, e_p=0.9, R=2, ρ_A=ρ_B). a) solids fraction profile. b) diffusion force profile. c) thermal diffusion factor profile.

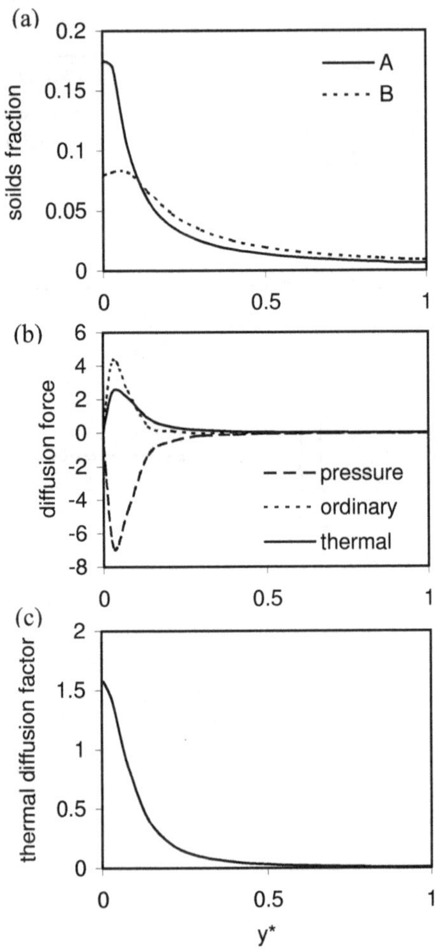

Fig 5.13: Comparison of flow profiles, for the JM model and IA model ($\bar{\phi}$=0.08, χ=1, e_p=0.9, R=4, M=1). a) solids fraction profile. b) granular energy profile. c) velocity profile.

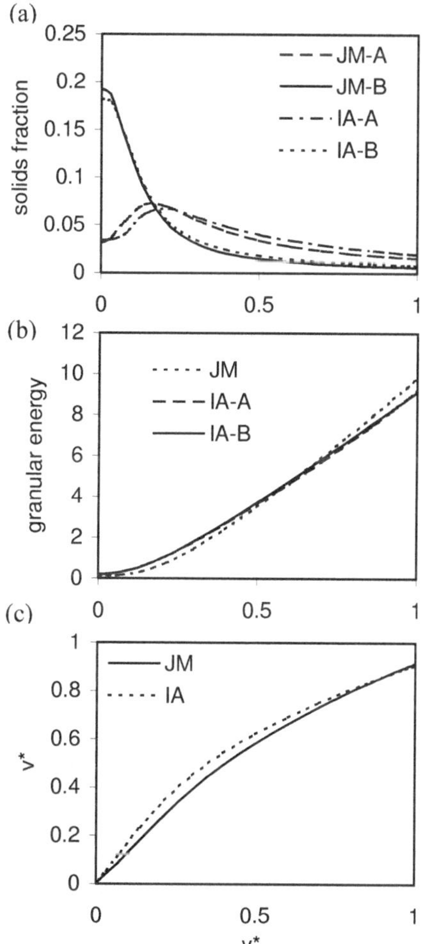

Fig 5.14: Diffusion profiles corresponding to the flow profiles in Fig 5.13 ($\bar{\phi}$=0.08, χ=1, e_p=0.9, R=4, M=1). a) diffusion force profile, for the JM model. b) diffusion force profile, for the IA model. c) thermal segregation profile.

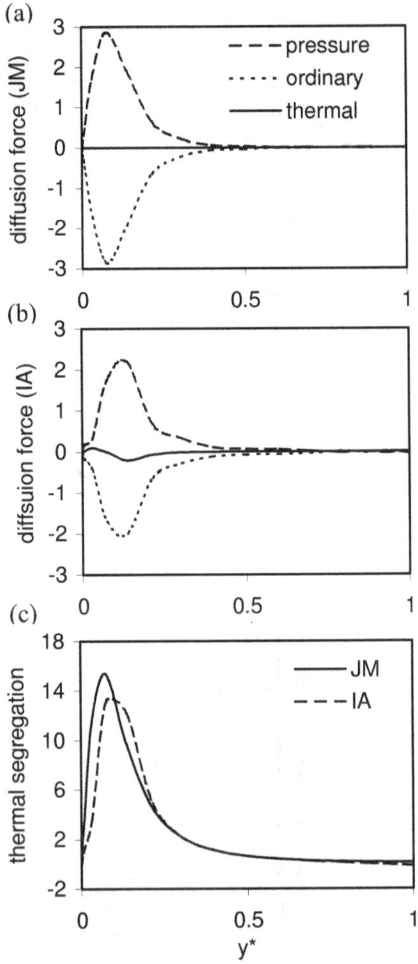

Fig 5.15: Transition of segregation with an increase in the mass ratio, based on the IA model ($\bar{\phi}$=0.05, χ=1, e_p=0.9, R=2). Solids fraction profile and corresponding diffusion force profile for a) M=1. b) M=1.5. c) M=2. d) M=2.5. e) M=4.

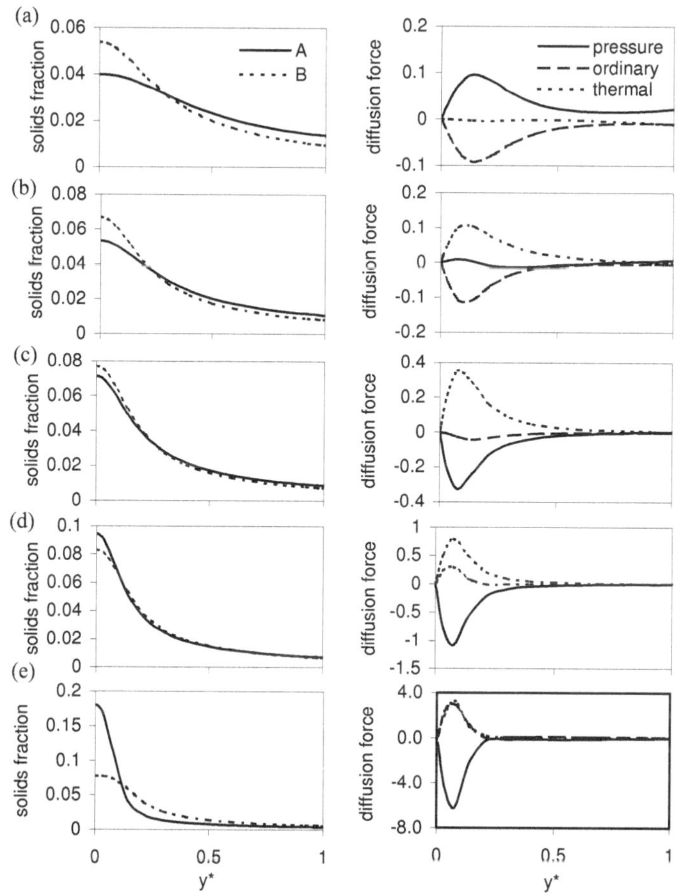

Fig 5.16: Effect of the size ratio on the average granular energy ($\bar{\phi}$=0.04, χ=1, e_p=0.9, M=1). a) variation of average granular energy with the size ratio. b) variation of granular energy ratio with the size ratio.

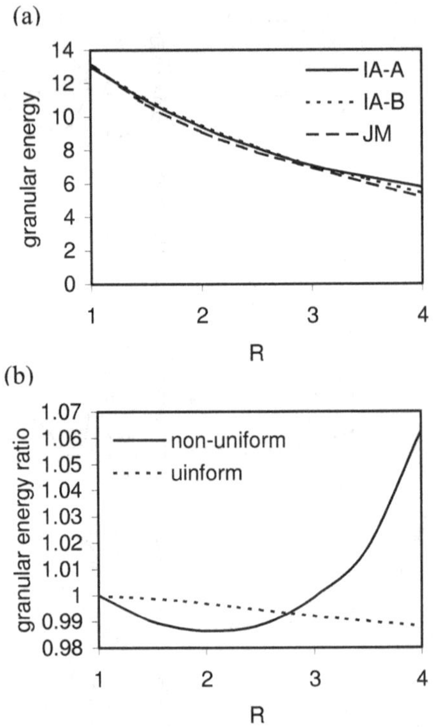

Fig 5.17: Effect of the mass ratio on the flow profiles for the JM model ($\bar{\phi}$=0.08, χ=1, e_p=0.9, R=1). a) solids fraction profile of large particles. b) solids fraction profile of small particles. c) granular energy profile. d) velocity profile.

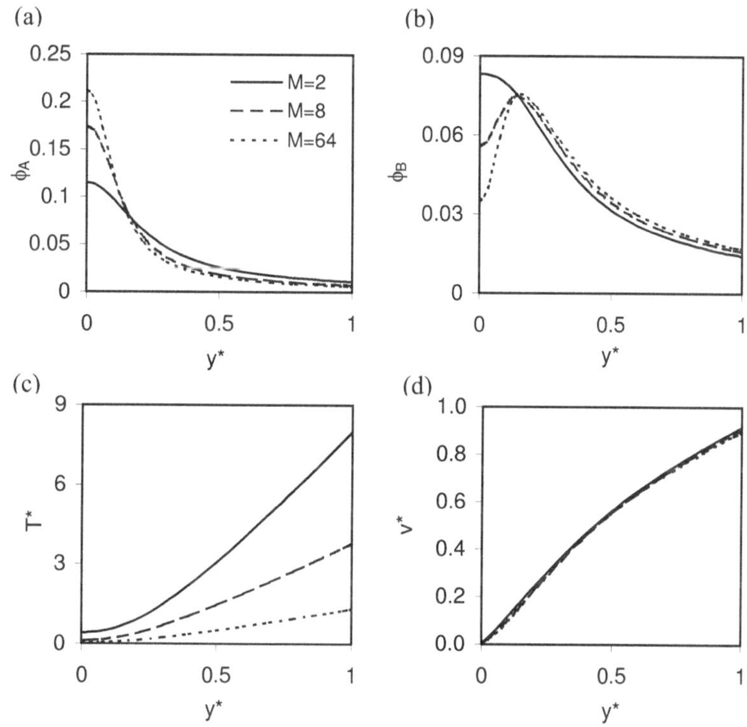

Fig 5.18: Effect of the mass ratio on the flow profiles for the IA model ($\bar{\phi}$=0.08, χ=1, e_p=0.9, R=1). a) solids fraction profile of large particles. b) solids fraction profile of small particles. c) granular energy profile of large particles. d) granular energy profile of small particles. d) velocity profile.

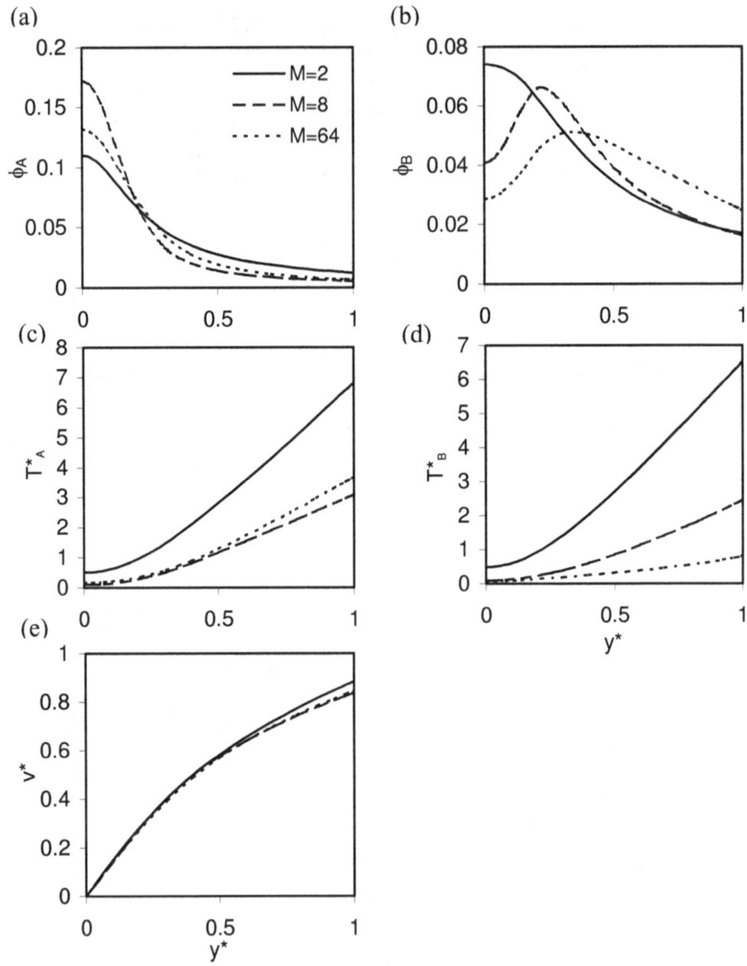

Fig 5.19: Effect of the mass ratio on the diffusion force profiles ($\bar{\phi}$=0.08, χ=1, e_p=0.9, R=1). a) diffusion force profile, for the JM model (M=8). b) diffusion force profile, for the JM model (M=64). c) diffusion force profile, for the IA model (M=8). d) diffusion force profile, for the IA model (M=64).

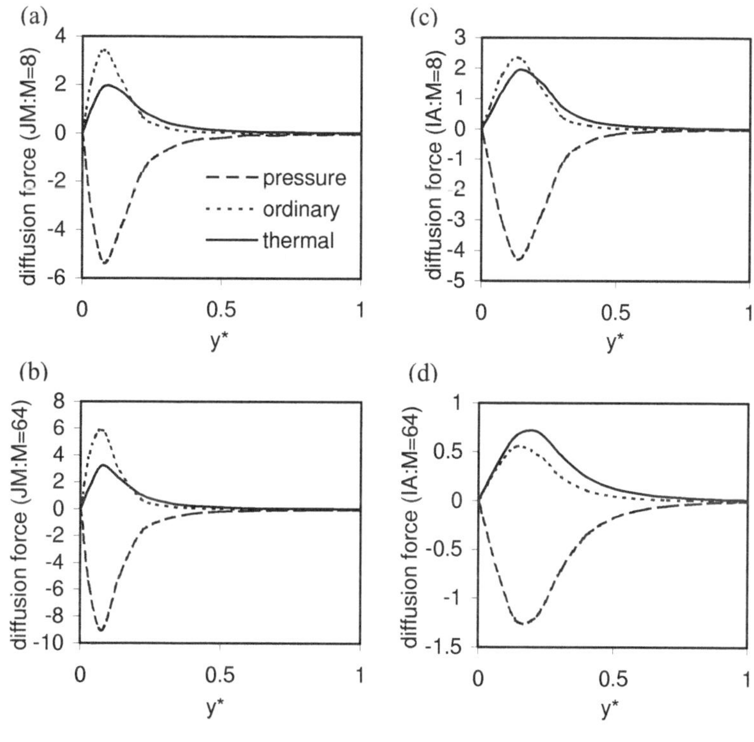

Fig 5.20: Effect of the mass ratio on the thermal segregation ($\bar{\phi}$=0.08, χ=1, e_p=0.9, R=1). a) thermal diffusion factor profile, for the JM model. b) thermal segregation profile, for the JM model. c) thermal diffusion factor profile, for the IA model. d) thermal segregation profile, for the IA model.

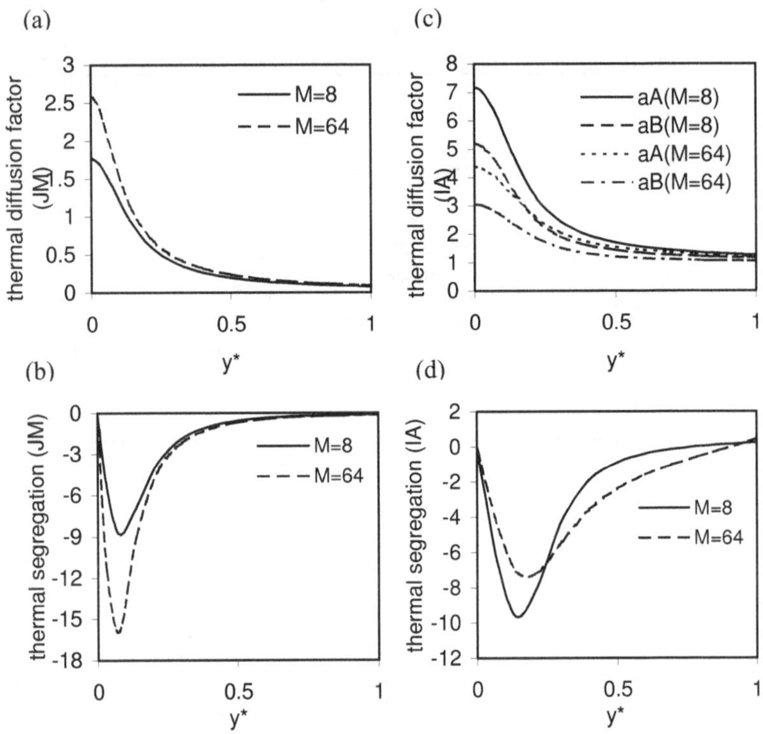

Fig 5.21: Effect of the mass ratio on the average granular energy ($\bar{\phi}$=0.07, χ=1, e_p=0.9, R=1). a) variation of average granular energy with the mass ratio. b) variation of granular energy ratio with the mass ratio.

(a)

(b)

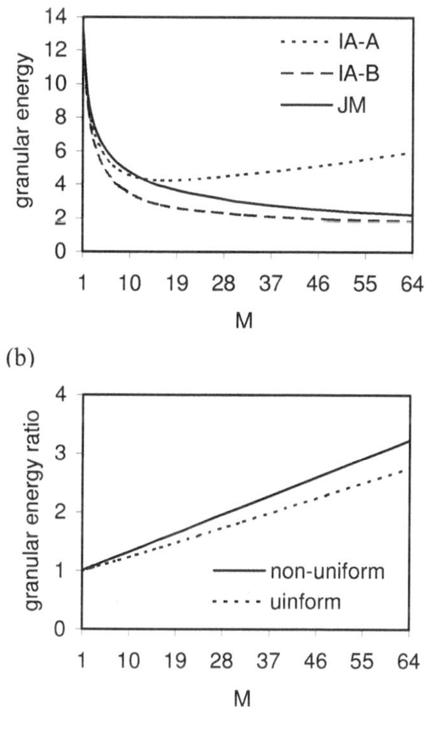

Fig 5.22: Granular channel flow profiles with a mono-disperse particle mixture ($\bar{\phi}$=0.1).

a) solids fraction profile. b) granular energy profile. c) velocity profile.

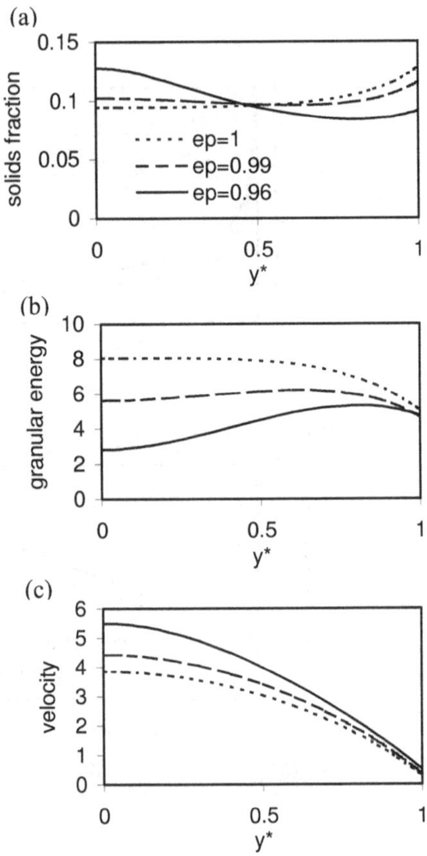

Fig 5.23: Granular channel flow profiles when e_p=1 ($\bar{\phi}$=0.1, $\chi=\bar{\phi}_A/\bar{\phi}_B$=1, R=$d_A/d_B$=2, $\rho_A=\rho_B$). a) solids fraction profile. b) granular energy profile. c) velocity profile.

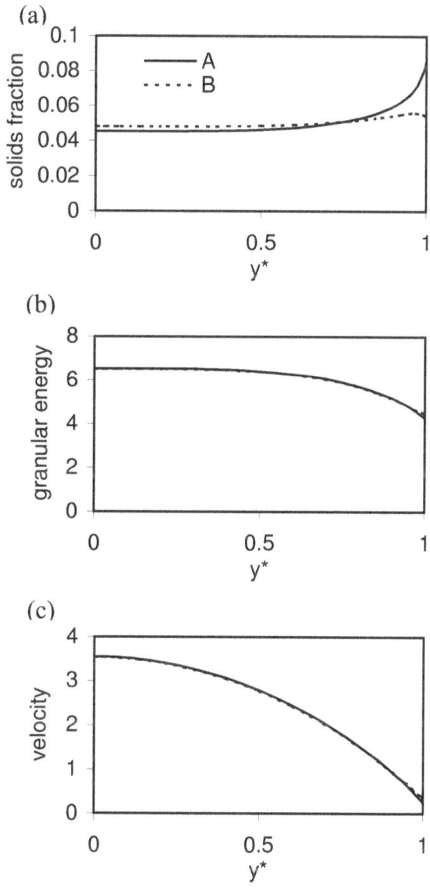

Fig 5.24: Diffusion profiles when e_p=1 ($\bar{\phi}$=0.1, χ=1, R=2, ρ_A=ρ_B). a) ordinary diffusion coefficient profile. b) diffusion force profile.

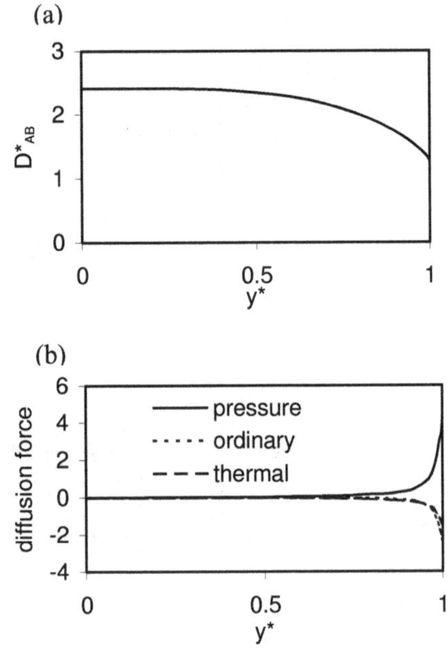

Fig 5.25: Granular channel flow profiles when e_p=0.99 ($\bar{\phi}$=0.1, χ=1, R=2, ρ_A=ρ_B). a) solids fraction profile. b) granular energy profile. c) velocity profile.

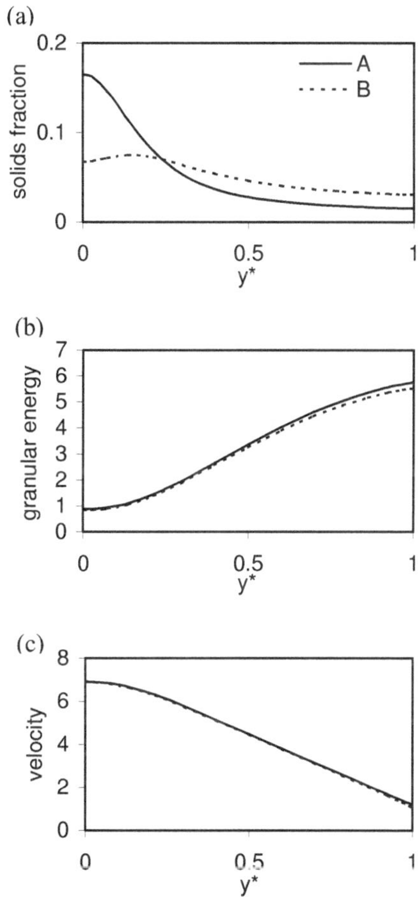

Fig 5.26: Diffusion profiles when e_p=0.99 ($\bar{\phi}$=0.1, χ=1, R=2, ρ_A=ρ_B). a) ordinary diffusion coefficient profile. b) diffusion force profile.

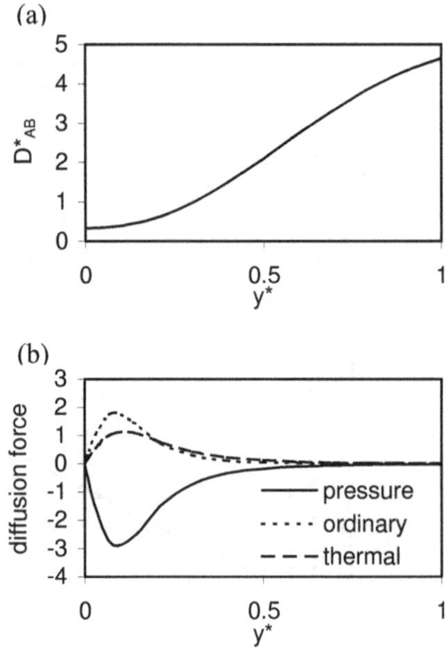

Fig 5.27: Granular channel flow profiles and diffusion force profiles when e_p=0.997 ($\bar{\phi}$=0.1, χ=1, R=2, ρ_A=ρ_B). a) solids fraction profile. b) granular energy profile. c) diffusion force profile.

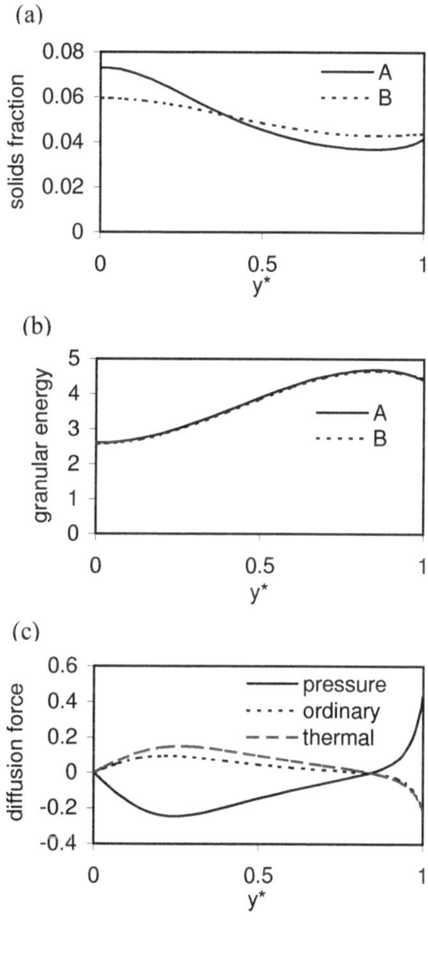

Fig 5.28: Granular channel flow profiles when e_p=0.96 ($\bar{\phi}$=0.1, χ=1, R=2, ρ_A=ρ_B). a) solids fraction profile. b) granular energy profile. c) velocity profile.

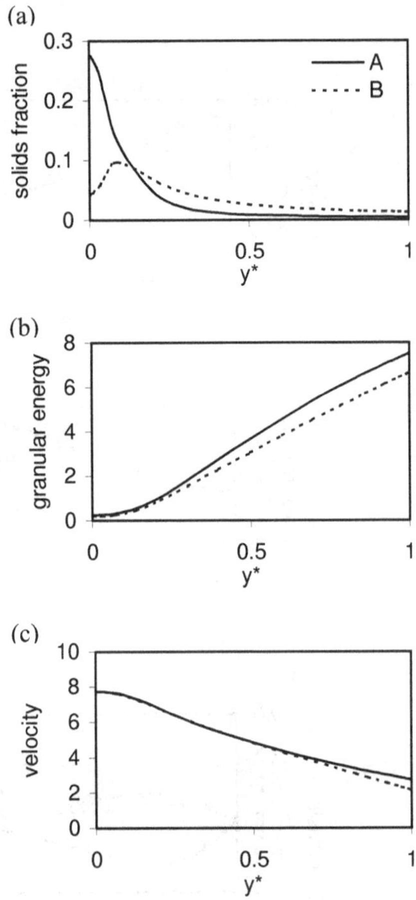

Fig 5.29: Effect of the coefficient of restitution on the total solids fraction profile ($\bar{\phi}$=0.1, χ=1). a) total solids fraction profile (R=d_A/d_B=2, $\rho_A=\rho_B$). b) total solids fraction profile (R=d_A/d_C=0.75, $\rho_A=\rho_C$).

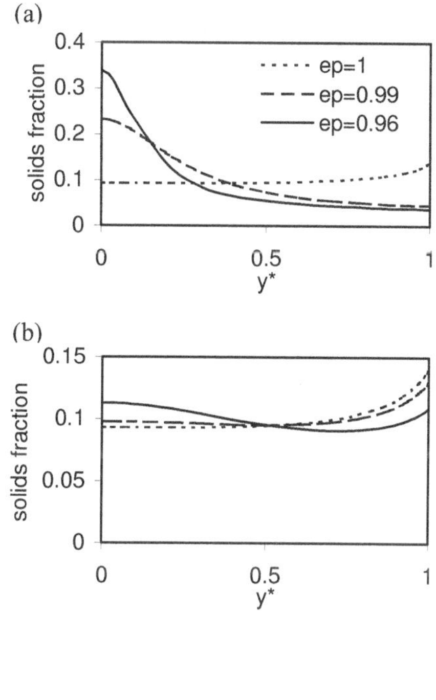

Fig 5.30: Solids fraction profiles and diffusion force profiles when $\bar{\phi}$=0.2 (χ=1, R=2, ρ_A=ρ_B). a) solids fraction profile (e_p=1). b) solids fraction profile (e_p=0.99). c) diffusion force profile (e_p=1). d) diffusion force profile (e_p=0.99).

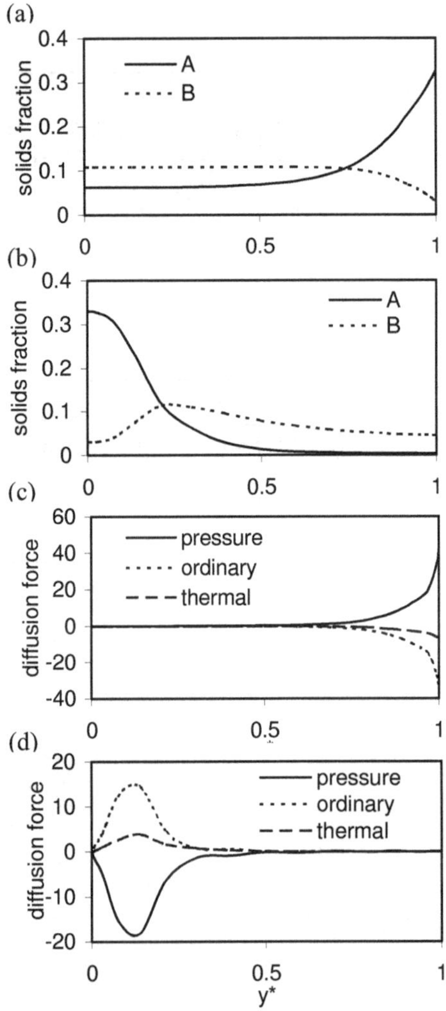

Fig 5.31: Effect of the size ratio on the granular channel flow profiles ($\bar{\phi}$=0.5, χ=1, e_p=0.99, $\rho_A=\rho_B$). a) solids fraction profile of large particles. b) solids fraction profile of small particles. c) granular energy profile of large particles. d) granular energy profile of small particles. e) velocity profile of large particles. f) velocity profile of small particles.

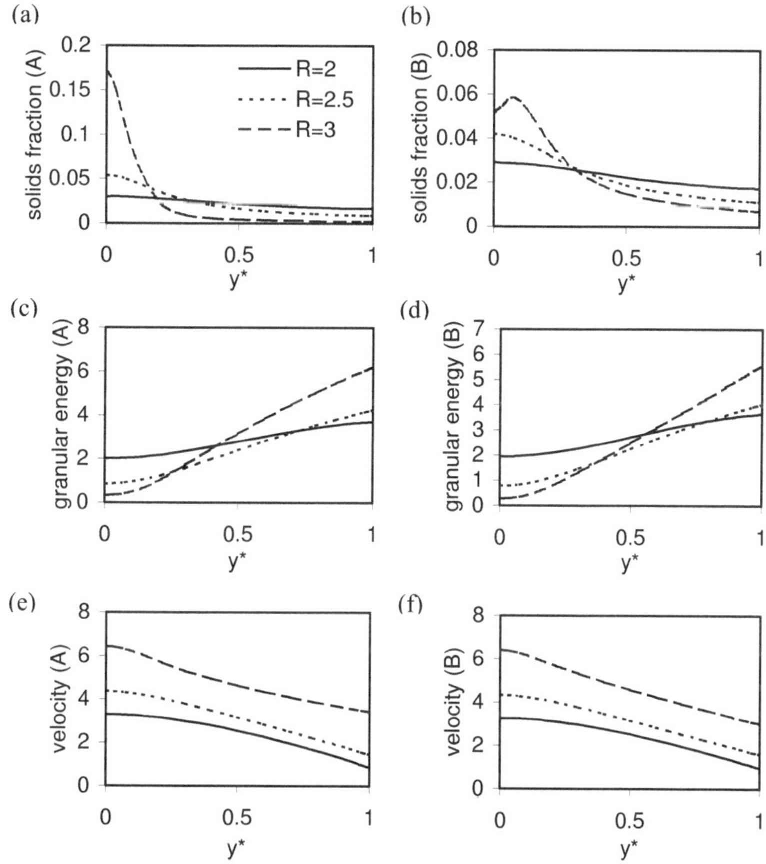

Fig 5.32: Effect of the size ratio on the granular flow profiles for the equal mass case ($\bar{\phi}$=0.1, χ=1, M=m_A/m_B=1, e_p=0.96). a) solids fraction profile of large particles. b) solids fraction profile of small particles. c) granular energy profile of large particles. d) granular energy profile of small particles. e) velocity profile of large particles. f) velocity profile of small particles.

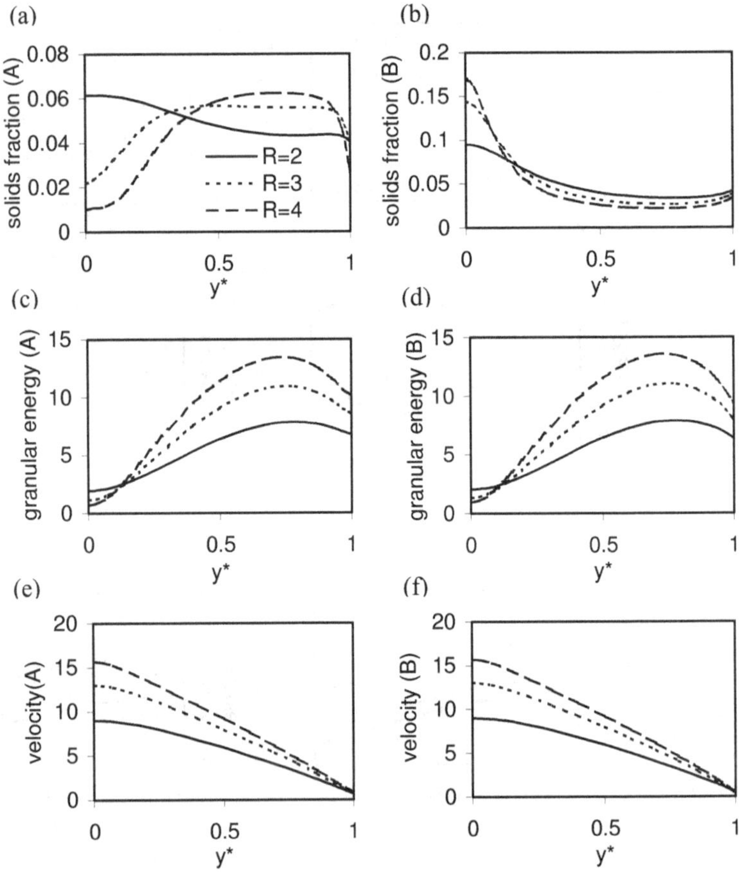

Fig 5.33: Effect of the size ratio on the diffusion profiles for the equal mass case ($\bar{\phi}$=0.1, χ=1, M=1, e_p=0.96). a) ordinary diffusion coefficient profile. b) pressure diffusion force (PDF) profile. c) ordinary diffusion force (ODF) profile. d) thermal diffusion force (TDF) profile.

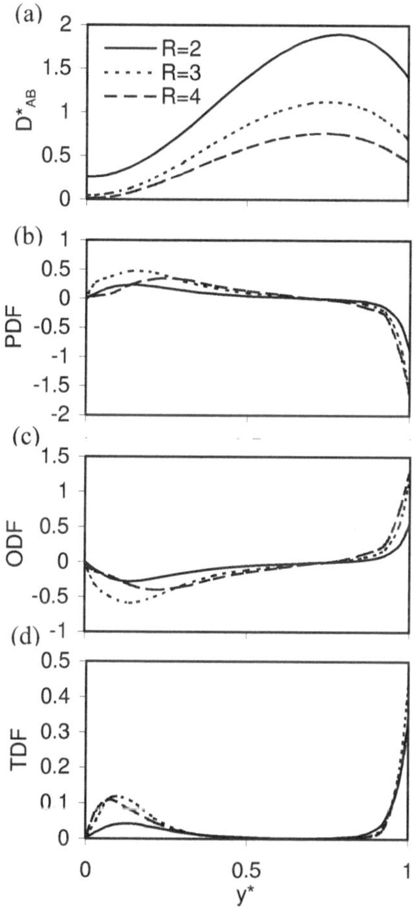

Fig 5.34: Effect of the mass ratio on the granular flow profiles for the equal size case ($\bar{\phi}$=0.1, χ=1, R=1, e_p=0.96). a) solids fraction profile of large particles. b) solids fraction profile of small particles. c) granular energy profile of large particles. d) granular energy profile of small particles. e) velocity profile of large particles. f) velocity profile of small particles.

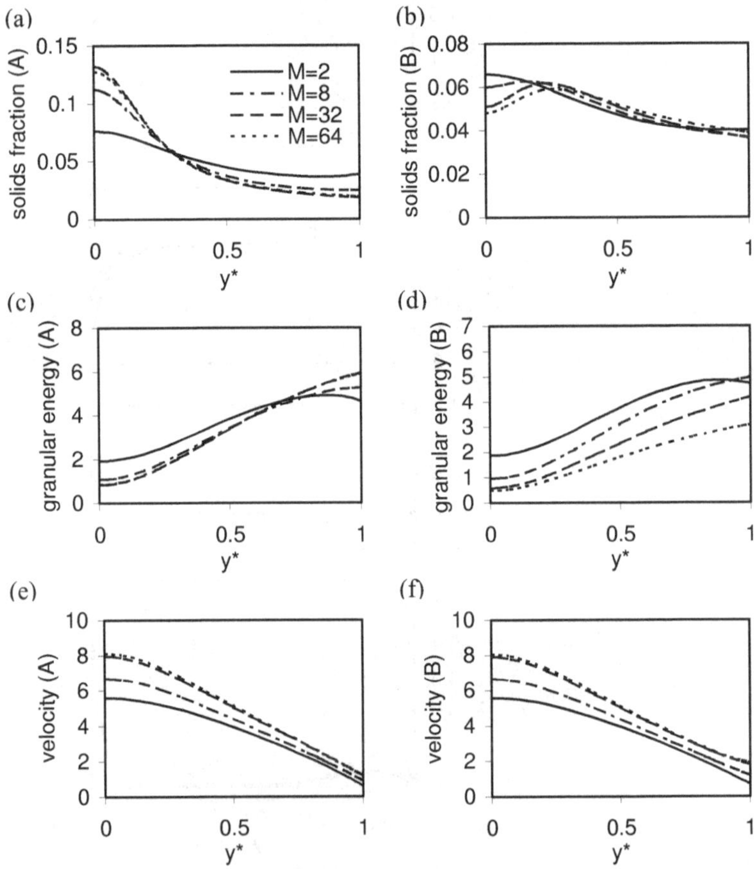

Fig 5.35: Effect of the mass ratio on the diffusion profiles for the equal size case ($\bar{\phi}$=0.1, χ=1, R=1, e_p=0.96). a) ordinary diffusion coefficient profile. b) pressure diffusion force profile. c) ordinary diffusion force profile. d) thermal diffusion force profile.

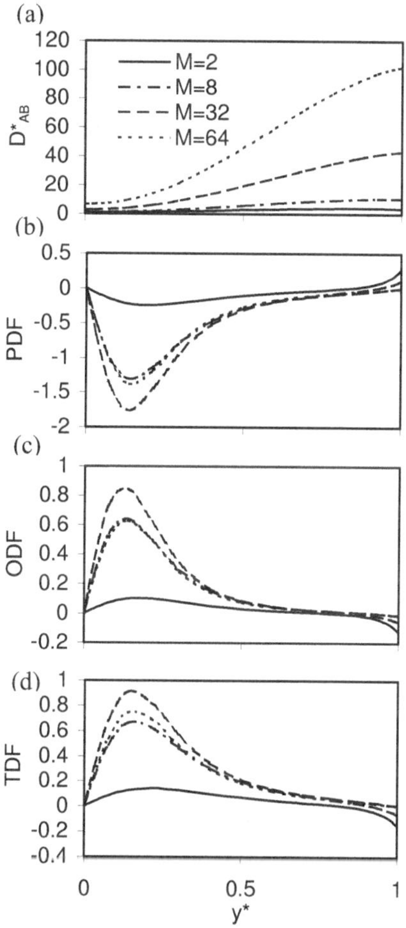

Fig 5.36: Gas-particle flow profiles and diffusion force profiles when e_p=1 ($\bar{\phi}$=0.1, χ=1, R=2, ρ_A=ρ_B). a) solids fraction profile. b) granular energy profile. c) velocity profile. d) diffusion force profile.

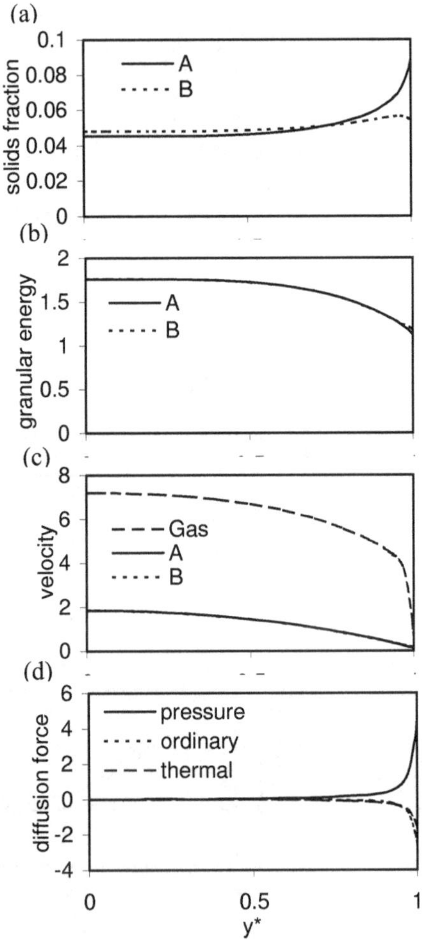

Fig 5.37: Gas-particle flow profiles and diffusion force profiles when e_p=0.99 ($\bar{\phi}$=0.1, χ=1, R=2, ρ_A=ρ_B). a) solids fraction profile. b) granular energy profile. c) velocity profile. d) diffusion force profile.

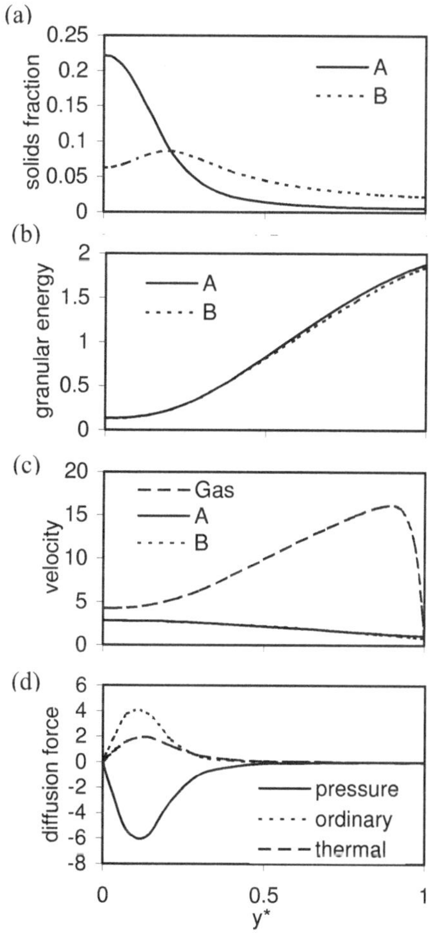

Chapter 6 Conclusions and Further Work

The rheology of granular materials is more complicated than that of fluids, and a qualitative and quantitative understanding of the flow behavior of particulate solids is a long-standing goal of engineers and scientists. Several specific challenges facing those involved in designing and analyzing particulate processes and gas-particle flows have been investigated numerically in this work.

We start by studying the impact of physical and process parameters on flow of gas and particles in a vertical duct with a mono-disperse particle mixture. Steady state solutions were sought to equations of continuity and motion with closures from kinetic theory and the variations of the solutions with changes in parameters were computed using a bifurcation/continuation approach. When the particles are assumed elastic, we find that the segregation of particles transits from the center to the walls with an increase in the bed width or solids load. The origin of this transition is due to the contribution of the gas-particle slip term on the pseudo-thermal energy transport. When the particles are inelastic, we find particles always segregate to the center of the channel, irrespective of the bed width or solids load. As we track the variation of the average solids fraction with a number of parameters, multiple steady state solutions can be observed. Recent interest in using fluidized beds on Mars or the Moon for local production of oxygen has prompted us to examine the effect of the gravity on steady fully developed flows in gas-particle fluidized beds. When the average gas velocity is held constant, the flow profiles are found to be insensitive to the magnitude of gravity if the gas flow rate is high and particles always move upward in the bed. The physical mechanism leading to this insensitivity can be traced to the automatic compensation of the gas-particle drag force to the body force. However, when the gas flow rate is low, particles move downward in the region close to the walls with high gravity while this downflow is not observed under low gravity. Finally, the effects of two dimensionless numbers are investigated. It is found that the flow behaviors are not sensitive to the variation of the Reynolds number when this number is high due to a relatively small value of the gas viscous term. With an increase in the Froude number, the gas and particle velocities decrease as well as the granular temperature.

Then we provide insight to the linear instability of gas-particle flows in a vertical duct. The stability analysis is carried out for three illustrative sets which correspond to walls being neglected, or acting as sinks or sources of fluctuational energy. Without considering the effect of walls, equations of motion have a solution representing a uniformly fluidized bed. The instability has the form of vertically traveling waves and its development is much faster than transverse structures. Using the current kinetic theory model, the uniform base state corresponding to walls being neglected is always unstable to small perturbations. With adding walls, the gas-fluidized bed system is more easily stabilized due to a much higher level of the granular temperature, and the non-uniform base state can remain stable or lose its stability to small perturbations, depending on the material properties and operational parameters. Generally, increasing solids fraction and decreasing inelasticity can suppress the instabilities. The dominant instability pattern has two types of structures: symmetric ones and anti-symmetric ones. Both of them propagate in the bed in the form of traveling waves at a speed comparable with that of the solids phase. From Fig 4.19 and 4.20, it is clear that different instability patterns can develop different density structures. Basically, symmetric patterns can develop layering structures and bubbles or clusters, and anti-symmetric patterns can develop streamer-like structures. This predictions can in principle be used to test the continuum models using experimental measurements. Experimental work has shown that for a high average solids fraction, the flow behavior shows bubbling structures, while for a low solids fraction streamer-like structures can be developed in the bed. The contourmap shown in Fig 4.11 provides a boundary for these two different flow situations. With a decrease in the solids fraction or restitution coefficient, the instability patterns transition between symmetric patterns and anti-symmetric patterns, leading to the transition of density structures between bubbling patterns and streamer patterns. Using this contourmap, we can compare our simulation predictions with experimental results if the experiments are designed to decrease the solids fraction or the restitution coefficient while holding other parameters unchanged. By comparing the leading eigenvalue profiles and dominant eigenfunction contour maps of sink wall conditions and source wall conditions, it is found that the stability behaviors are not sensitive to the physical nature of the boundary walls due to the similarity of their base state solids fraction profiles. The physical mechanism of the instabilities is investigated using a term-by-term analysis. All instabilities reported in this work are suppressed if solid phase inertia or inelastic particle collisions are removed. While previous studies have neglected gas phase inertia, we show that for our conditions these terms can have an important contribution. In particular, we show that the development of symmetric patterns occurs due to the contribution of the gas phase inertial terms. For the chain reaction studied in the present work, we find that the flow behaviors and instability natures strongly affect the product yield and selectivity. For steady state flows, corresponding to the non-uniform distribution of the flow profiles, the concentration profile shows a large fluctuation across the bed (see Fig 4.32 and 4.33). When the development of the density structures is dominated by symmetric or anti-symmetric patterns, the final distribution of products is completely different to that for the steady state flows, indicating the instabilities have a significant effect on the chemical reactions and the concentration distribution of the final products.

To study the effect of poly-disperse particles, we extend our work to investigate the steady state solutions of boundary-driven and body-force-driven granular flows and gas-particle channel flows with a bi-disperse particle mixture using kinetic theory. For granular flow, we found that energy equipartition breaks down with an increase in the system inelasticity and the mass ratio, whereas the effect of the size ratio is very small if the two particle species have the same mass. Two forms of segregation are studied in the present work: the total solids segregation in the system and the solids species segregation. The species segregation is enhanced with a decrease in the system inelasticity, an increase in the average solids fraction or an increase in the size ratio. The total solids segregation is attributed to the distribution of the granular energy since particles have a potential to move to low energy regions. The solids species segregation is due to the competition of three diffusion forces: the thermal diffusion force, the ordinary diffusion force and the pressure diffusion force. The segregation of total particles and each individual particle species is enhanced when the effect of non-equipartition of granular energy is considered. For equal mass particles, small particles have a higher solids fraction in low energy regions. For equal size particles the solids fraction of heavy particles is higher in low energy regions. The segregation in the equal density mixture is due to the competition of mass disparity and size disparity. For the equal density case, large particles have more potential to segregate in low granular energy regions indicating that the contribution of the mass disparity is more significant than the effect of the size disparity. For body-force-driven granular flows, the segregation of large particles has a transition from the walls to the center when e_p decreases from 1 to 0.99. This sensitivity is reduced when the system becomes more inelastic. By comparing with the monodisperse system, we show that if larger particles are mixed in the system the sensitivity of the total solids distribution to the restitution coefficient is suppressed, while if smaller particles are added in the system the situation reverses. Finally we extend our work to gas-particle flows in a channel, and find that for the kinetic theory models used in the present study, the solids fraction and the granular temperature profiles are quite similar between the granular flows and the gas-particle flows, indicating that the effect of the gas is mainly in the vertical direction, which drive the particles to be suspended in the bed.

Following the results in this dissertation, some of the interesting directions to pursue in the future are:

(1) Potential improvement of the current model can be carried out by considering the following points: the frictional contribution to the stress tensor becomes important when the solids fraction is high (>0.5) and particles are in sustained contact with one another under this situation; the granular materials used in the present work are assumed to be cohesionless while many practical applications involve the flow of cohesive particles; the contribution of the particle rotation and spin is not considered in the present model. Further advances will hinge on the incorporation of these factors in the model. In addition, it is not fully understood how sensitive the predictions from a continuum model

are to the choices of the constitutive relations, such as solids stress and pseudothermal energy flux. A further investigation of different kinetic theory models is necessary.

(2) In the present work, only the linear contribution is considered for the stability test (see Chapter 4). A main assumption in the results of Fig 4.19 and Fig 4.20 is that the short temporal evolution can be governed by the linear instabilities. It is of interest to examine how these waves grow in the long term and if the patterns developed have any resemblance to the dominant instability patterns predicted by the linear theory. These questions can be answered by a transient development approach. It is of interest to extend the transient development theory to the gas-fluidized bed system to see how the density waves, obtained in the present work, develop temporally and whether nonlinear terms are significant in the transient integration.

(3) A long term goal of work on gas-particle fluidized beds is combining the flow hydrodynamics and the chemical reactions together to study how the flow behaviors and instabilities affect the chemical reactions. In the present work, a chain reaction is assumed to ensure that the chemical reaction does not affect the flow behaviors, and thus we can calculate the flow hydrodynamics and reactions separately (see Chapter 4). It is of interest to extend our work to real chemical reaction systems.

(4) For poly-disperse systems, only steady state solutions are investigated in the current work (see Chapter 5). To see whether the base state is stable, a linear stability analysis is necessary. It is also of interest to perform direct integration of the equations of motion to study the transient behavior in bidisperse particle mixtures. We can discretize the system spatially and specify two initial conditions, corresponding to a well-mixed state and a fully-segregated state, respectively. Then the mixing and segregation behaviors can be investigated when simulating the particulate system temporally.

References

Abramowitz, M., and I. A. Stegun, 1965, Handbook of Mathematical Function: Appl. Math. Ser., v. 55.

Agbim, J. A., A. W. Nienow, and P. N. Rowe, 1971, Inter-particle forces that suppress bubbling in gas-fluidized beds: Chem. Eng. Sci., v. 26, p. 1293-1294.

Agrawal, K., P. Loezos, M. Syamlal, and S. Sundaresan, 2001, The role of meso-scale structures in rapid gas–solid flows: J. Fluid Mech., v. 445, p. 151-185.

Alam, M., and S. Luding, 2002a, How good is the equipartition assumption for the transport properties of a granular mixture: Granular Matter, v. 4, p. 139-142.

Alam, M., and S. Luding, 2002b, Kinetic theory of a binary mixture of nearly elastic disks with size and mass disparity: Phys. Fluids, v. 14, p. 4085-4087.

Alam, M., and S. Luding, 2003, Rheology of bidisperse granular mixtures via event-driven simulations: J. Fluid Mech., v. 476, p. 69-103.

Alam, M., and P. R. Nott, 1998, Stability of plane couette flow of a granular material: J. Fluid Mech., v. 377, p. 99-136.

Anderson, K., S. Sundaresan, and R. Jackson, 1995, Instabilities and the formation of bubbles in fluidized beds: J. Fluid Mech., v. 303, p. 327-366.

Anderson, T. B., and R. Jackson, 1967, A fluid mechanical description of fluidized beds. Equations of motion: Ind. Eng. Chem. Fundam., v. 6, p. 527-539.

Anderson, T. B., and R. Jackson, 1968, Fluid mechanical description of fluidized beds. Stability of the state of uniform fluidization: Ind. Eng. Chem. Fundam., v. 7, p. 12-21.

Anderson, T. B., and R. Jackson, 1969, Fluid mechanical description of fluidized beds. Comparsion of theory and experiment: Ind. Eng. Chem. Fundam., v. 8, p. 137-144.

Arastoopour, H., S. C. Lin, and S. A. Weil, 1982, Analysis of vertical pneumatic conveying of solids using multiphase flow models: AICHE J., v. 28, p. 467-473.

Arnarson, B. O., and J. T. Willits, 1998, Thermal diffusion in binary mixtures of smooth, nearly elastic spheres with and withour gravity: Phys. Fluids, v. 10, p. 1324-1328.

Bachelor, G. K., 1993, Secondary instability of a gas-fluidized bed: J. Fluid Mech., v. 257, p. 359-371.

Bachelor, G. K., and J. M. Nitsche, 1991, Instability of stationary unbounded stratified fluid: J. Fluid Mech., v. 227, p. 357-391.

Bader, R., J. Findlay, and T. Knowlton, 1988, Circulating Fluidized Bed Technology II, Pergamon, New York.

Bagnold, R. A., 1954, Experiments on a gravity-free dispersion of large solid spheres in a newtonian fluid under shear: Proc. R. Soc. Lond. A., v. 225, p. 49-63.

Barrat, A., and E. Trizac, 2002, Lack of energy equipartition in homogeneous heated binary granular mixtures: Granular Matter, v. 4, p. 57-63.

Bellomo, N., and M. Pulvirenti, 2000, Modeling in applied science s a kinetic theory approach. Modeling and Simulation in Science and Technology.

Benyahia, A., H. Arastoopour, T. M. Knowlton, and H. Massah, 2000, Simulation of particles and gas flow behavior in the riser section of a circulating fluidized bed

using the kinetic theory approach for the particulate phase: Powder Technology, v. 112, p. 24-33.
Bird, R. B., W. E. Stewart, and E. N. Lightfoot, 1960, Transport Phenomena, John Wiley & Sons, Inc.
Bridgewater, J., 1976, Fundamental powder mixing mechanisms: Powder Technology, v. 15, p. 215-236.
Bridgewater, J., 2003, Granular matter: Towards understanding the fundamentals.: Granular matter.
Bridgewater, J., M. H. Cooke, and A. M. Scott, 1978, Inter-particle percolation: Equipment development and mean percolation velocities: Trans. Inst. Chem. Eng., v. 56, p. 157-167.
Bridgewater, J., W. S. Foo, and D. J. Stephens, 1985, Particle mixing and segregation in failure zones - theory and experiment.: Powder Technology, v. 41, p. 147-158.
Chapman, S., and T. G. Cowling, 1970, The Mathematical Theory of Non-uniform Gases.
Cody, G. D., R. J. Bellows, D. J. Goldfarb, H. A. Wolf, and G. V. Storch, Jr .. 2000, Anovel non-intrusive probe of particle motion and gas generation in the feed injection zone of the feed riser of a fluidized catalytic cracking unit: Powder Technology, v. 110, p. 128-142.
Cody, G. D., D. J. Goldfarb, G. V. J. Storch, and A. N. Norris, 1996, Particle granular temperature in gas fluidized beds: Powder Technology, v. 87, p. 211-232.
Constantinides, A., 1987, Chemical Engineering Series - Applied Numerical Methods with Personal Computers, McGraw-Hill Inc.
Conway, S. L., and B. J. Glasser, 2004, Density Waves and Coherent Structures in Granular Couette Flows: Phys. Fluids, v. 16, p. 509-529.
Cross, M. C., and P. C. Hohenburg, 1993, Pattern formation outside of equilibrium: Reviews of modern physics, v. 65, p. 851-1172.
Dankworth, D. C., and S. Sundaresan, 1991, Time-dependent flow patterns arising from the instability of uniform fluidization: Presentation at Symposium on Mechanics of Fluidized Beds, Stanford University.
Dasgupta, S., R. Jackson, and S. Sundaresan, 1994, Turbulent gas-particle flow in vertical risers: AICHE J., v. 40, p. 215-228.
Dasgupta, S., R. Jackson, and S. Sundaresan, 1998, Gas-particle flow in vertical pipes with high mass loading of particles: Powder Technology, v. 116, p. 190-203.
Davidson, J. F., R. Clift, and D. Harrison, 1985, Fluidization, 2nd edition.
Detamore, M. S., M. A. Swanson, K. R. Frender, and C. M. Hrenya, 2000, A kinetic-theory analysis of the scale-up of circulating fluidized beds: Powder Technology, v. 110, p. 210-221.
Deuflhard, P., E. Hairer, and J. Zugck, 1987, One-step and extrapolation methods for differential-algebraic systems: Numer. Math, v. 51, p. 501-516.
Didwania, A. K., and G. M. Homsy, 1981, Flow regimes and flow transitions in liquid fluidized beds: International Journal of Multiphase Flow, v. 7, p. 563-580.
Ding, J., and D. Gidaspow, 1990, A bubbling fluidization model using kinetic theory of granular flow: AICHE J., v. 36, p. 523.

Doedel, E., H. Keller, and J. Kernevez, 1991, Numerical analysis and control of bifurcation problems. I: Bifurcation in finite dimensions: Intl. J. Bifurcation Chaos, v. 1, p. 493-520.
Dolgunin, V. N., A. N. Kudy, and A. A. Ukolov, 1998, Development of the model of segregation of particles undergoing granular flow down an inclined chute: Powder Technology, v. 96, p. 211-218.
Drew, D. A., 1971, Averaged field equations for two-phase media: Stud. Appl. Math, v. 50, p. 133-166.
Eduardo, J. B., J. A. Yasuna, and J. L. Sinclair, 1995, Dilute turbulent gas-solid flow in risers with particle-particle interactions: AICHE J., v. 41, p. 1375-1388.
Ehrig, R., U. Nowak, L. Oeverdieck, and P. Deuflhard, 1999, High performance scientific and engineering computing: Bungartz, H. J., Durst, F. and Zenger, C. (eds.), Lecture, Springer, v. 8, p. 233.
El-Kaissy, M. M., and G. M. Homsy, 1976, Instability waves and the origin of bubbles in fluidized beds. I: Experiments: International Journal of Multiphase Flow, v. 2, p. 379-395.
Fan, Y. P., and Z. Chao, 1998, Principles of Gas-Solids Flows, Cambridge University Press.
Fan, Y. P., S. Ye, Z. X. Chao, C. X. Lu, G. G. Sun, and M. X. Shi, 2002, Gas-solid two-phase flow in FCC riser: AICHE J., v. 48, p. 1869-1887.
Fanucci, J. B., N. Ness, and R. H. Yen, 1979, On the formation of bubbles in gas-particle fluidized beds: J. Fluid Mech., v. 94, p. 353-367.
Feitosa, K., and N. Menon, 2002, Breakdown of energy equipartition in a 2D binary vebrated granular gas: Phys. Rev. Lett., v. 88, p. 198301.
Galvin, J. E., S. R. Dahl, and C. M. Hrenya, 2005, On the role of non-equipartition in the dynamics of rapidly-flowing, granular mixtures: J. Fluid Mech., v. 528, p. 207-232.
Garzo, V., and J. Dufty, 1999, Homogeneous cooling state for a granular mixture: Physical Review E, v. 60, p. 5706-5713.
Geldart, D., 1973, Types of gas fluidization: Powder Technology, v. 7, p. 285-292.
Geldart, D., 1986, Gas Fluidization Technology.
Geldart, D., N. Harnby, and A. C. Wong, 1984, Fludization of conhesive powders: Powder Technology, v. 37, p. 25-37.
Geldart, D., and A. C. Wong, 1987: AICHE Symp. Ser., v. 83, p. 255.
Gibilaro, L. G., R. Foscolo, and D. F. U., 1986, The influence of gravity on the stability of fluidized beds: Chem. Eng. Sci., v. 41, p. 2438-2440.
Gidaspow, D., M. Syamlal, and Y. C. Seo, 1986, Hydrodynamics of fluidization: supercomputer generated vs. experimental bubbles: J. Powder and Bulk Solids Tech., v. 10, p. 19-23.
Glasser, B. J., I. G. Kevrekidis, and S. Sundaresan, 1996, One- and two-dimensional travelling wave solutiuons in gas fluidized beds: J. Fluid Mech., v. 306, p. 183-221.
Glasser, B. J., I. G. Kevrekidis, and S. Sundaresan, 1997, Fully developed travelling wave solutions and bubble formation in fluidized beds: J. Fluid Mech., v. 334, p. 157-188.

Glasser, B. J., S. Sundaresan, and I. G. Kevrekidism, 1998, From bubbles to clusters in fluidized beds: Phys. Rev. Lett., v. 81, p. 1849-1852.

Goldhirsch, I., 1999, Scales and kinetics of granular flows: Chaos, v. 9, p. 659-672.

Goldhirsch, I., and G. Zanetti, 1993, Clustering instability in dissipative gases: Phys. Rev. Lett., v. 70, p. 1619-1622.

Goldschmidt, M. J. V., K. J. A. M., and S. W. P. M. V., 2001, Hydrodynamic modelling of dense gas-fluidised beds using the kinetic theory of granular flow: effect of coefficient of restitution on bed dynamics: Chem. Eng. Sci., v. 56, p. 571-578.

Goldshtein, A., and M. Shapiro, 1995, Mechanics of collisional motion of granular materials. Part 1: General hydrodynamic equations: J. Fluid Mech., v. 282, p. 75-114.

Goz, M. F., 1992, On the origin of wave patterns in fluidized beds: J. Fluid Mech., v. 240, p. 379-404.

Goz, M. F., 1993, Bifurcation of plane voidage waves in fluidized beds: Physica D, v. 65, p. 319-351.

Goz, M. F., 1995, Transverse instability of plane wavetrains in gas-fluidized beds: J. Fluid Mech., v. 303, p. 55-81.

Grace, J. R., and G. Sun, 1991, Influence of particle size distribution on the performance of fluidized bed reactors: Can. J. Chem. Eng., v. 69, p. 1126.

Grag, S. K., and J. W. Pritchett, 1975, Dynamics of gas fluidized beds: J. Appl. Phys., p. 4493.

Hayakawa, H., T. S. Komatsu, and T. Tsuzuki, 1994, Pseudo-solutions in fluidized beds: Physica A, v. 204, p. 277-289.

Helland, E., R. Occelli, and L. Tadrist, 2000, Numerical study of cluster formation in a gas-particle circulating fluidized bed: Powder Technology, v. 110, p. 210-221.

Hirayama, O., and R. Takaki, 1997, Analysis of nonlinear waves in a one-dimensional fluidized bed: Fluid Dyn. Res., v. 21, p. 233-247.

Hirayama, O., and T. Tsuzuki, 1995, Linear stability of a two-dimensional uniform fluidized beds: Fluid Dyn. Res., v. 15, p. 425-435.

Homsy, G. M., 1998, Nonlinear waves and the origin of bubbles in fluidized beds: Applied Scientific Research, v. 58, p. 251-274.

Hoomans, B. P. B., J. A. M. Kuipers, M. A. Mohd Salleh, M. Stein, and J. P. K. Seville, 2001, Experimental validation of granular dynamics simulations of gas-fluidized beds with homogenous in-flow conditions using positron emission particle tracking: Powder Technology, v. 116, p. 166-177.

Howley, M. A., and B. J. Glasser, 2002, Hydrodynamics of a uniform liquid-fluidized bed containing a binary mixture of particles: Chem. Eng. Sci., v. 57, p. 4209-4226.

Hrenya, C. M., and J. L. Sinclair, 1997, Effects of particle-phase turbulence in gas-solid flows: AICHE J., v. 43, p. 853-869.

Hui, K., P. K. Haff, J. E. Ungar, and R. Jackson, 1984, Boundary conditions for high-shear grain flows: J. Fluid Mech., v. 145, p. 223-233.

Huilin, L., D. Gidaspow, and E. Manger, 2001, Kinetic theory of fluidized binary granular mixtures: Physical Review E, v. 64, p. 1-8.

Huilin, L., L. Wenti, B. Rushan, Y. Lidan, and D. Gidaspow, 2000, Kinetic theory of fluidized binary granular mixtures with unequal granular temperature: Phyica A, v. 284, p. 265-276.

Huilin, L., H. Yurong, and D. Gidaspow, 2003, Hydrodynamic modeling of binary mixture in a gas bubbling fluidized bed using the kinetic theory of granular flow: Chem. Eng. Sci., v. 58, p. 1197-1205.

Iddir, H., and H. Arastoopour, 2005, Modeling of multitype particle flow using the kinetic theory approach: AICHE J., v. 51, p. 1620-1632.

Jackson, R., 1963, the mechanics of fluidized beds. I: the stability of the state of uniform fluidization: Trans. Inst. Chem. Eng., v. 41, p. 13-21.

Jackson, R., 1971, Fluid mechanical theory. In Fluidization (Ed. Davidson, J. F. and Harrison, D.), p. 65-119.

Jackson, R., 1997, Locally averaged equations of motion for a mixture of identical spherical particles and a Newtonian fluid: Chem. Eng. Sci., v. 52, p. 2457-2469.

Jackson, R., 2000, The Dynamics of Fluidized Particles, Cambridge University Press.

Jayati, J., and B. J. Glasser, 2002, Connections between density waves in fluidized beds and compressible flows: AICHE J., v. 48, p. 1645-1664.

Jenkins, J. T., and F. Mancini, 1987, Balance laws and constitutive relations for plane flows of a dense, binary mixture of smooth, nearly elastic circular disks: J. Appl. Mech, v. 54, p. 27-34.

Jenkins, J. T., and F. Mancini, 1989, Kinetic theory for binary mixtures of smooth, nearly elastic spheres: Phys. Fluids A, v. 1, p. 2050-2057.

Jenkins, J. T., and M. Richman, 1985, Kinetic theory for plane flows of a dense gas of identical, rough, inelastic, circular disks: Phys. Fluids, v. 28, p. 3485-3494.

Jenkins, J. T., and M. W. Richman, 1986, Boundary conditions for plane flows of smooth, nearly elastic, circular disks: J. Fluid Mech., v. 171, p. 53-69.

Jenkins, J. T., and S. B. Savage, 1983, A theory for the rapid flow of identical, smooth, nearly elastic, spherical particles: J. Fluid Mech., v. 130, p. 187-202.

Johnson, P. C., and R. Jackson, 1987, Frictional-collisional constitutive relations for granular materials, with application to plane shearing: J. Fluid Mech., v. 176, p. 67-93.

Johnson, P. C., P. Nott, and R. Jackson, 1990, Frictional-collisional constitutive equations of motion for particulate flows and their application to chutes: J. Fluid Mech., v. 210, p. 501-535.

Khakhar, D. V., J. J. McCarthy, and J. M. Ottino, 1999, Mixing and segregation of granular materials in chute flows: Chaos, v. 9, p. 594-610.

Khinast, J. G., 2001, Impact of 2-D bubble dynamics on the selcetivity of fast gas-liquid reactions: AICHE, v. 47, p. 2304-2319.

Kincaid, J. M., E. G. D. Cohen, and d. H. M. Lopez, 1987, The Enskog theory for multicomponent mixtures . IV. Thermal diffusion: J. Chem. Phys., v. 86, p. 963.

Koch, D. L. S., A. S., 1999, Particle pressure and marginal stability limits for a homogeneous monodisperse gas fluidized bed: kinetic theory and mumerical simulation: J. Fluid Mech., v. 400, p. 229-263.

Koh, C. J., P. Hookham, and L. G. Leal, 1994, An experimental investigation of concentrated suspension flows in a rectangular channel: J. Fluid Mech., v. 266, p. 1-32.

Kuipers, J. A. M., W. Prins, and W. P. M. Van Swaaij, 1991, Theoretical and experimental bubble formation at a single orifice in a two-dimensional gas-fluidized bed: Chem. Eng. Sci., v. 46, p. 2881-2894.
Kunii, D., and O. Levenspiel, 1991, Fluidization Engineering, Butterworth-Heinemann.
Li, J., and J. A. M. Kuipers, 2003, Gas-particle interaction in dense gas-fluidized beds: Chem. Eng. Sci., v. 58, p. 711-718.
Liss, E. D., and B. J. Glasser, 2001, The influence of clusters on the stress in a sheared granular material: Powder Technology, v. 116, p. 116-132.
Liu, J. T. C., 1983, Nonlinear unstable wave disturbances in fluidized beds: Proc. R. Soc. Lond. A., v. 389, p. 331.
Liu, X., and B. J. Glasser, 2006, A parametric investigation of gas-particle flow in a vertical duct: AICHE, v. 52, p. 940-956.
Louge, M. Y., E. Mastorakos, and J. T. Jenkins, 1991, The role of particle collisions in pneumatic transport: J. Fluid Mech., v. 231, p. 345-359.
Lun, C. K. K., 1991, Kinetic theory for granular flow of dense, slightly inelastic, slightly rough spheres: J. Fluid Mech., v. 233, p. 539-559.
Lun, C. K. K., and S. B. Savage, 1986, A simple kinetic theory for granular flow of binary mixtures of smooth, inelastic, spherical particles: Acta Mehcanica, v. 63, p. 45-60.
Lun, C. K. K., and S. B. Savage, 1987, A simple kinetic theory for granular flow of rough, inelastic, spherical particles: J. Appl. Mech, v. 54, p. 47-53.
Lun, C. K. K., S. B. Savage, D. J. Jeffrey, and N. Chepurniy, 1984, Kinetic theories for granular flow: inelastic particles in couette flow and slightly inelastic particles in a general flow field: J. Fluid Mech., v. 140.
Lyon, M. K., and L. G. Leal, 1998, An experimental study of the motion of concentrated suspensions in two-dimensional channel flow. Part 1. Monodisperse systems: J. Fluid Mech., v. 363, p. 25-56.
Makkawi, Y. T., and P. C. Wright, 2003, the voidage function and effective drag force for fluidized beds: Chem. Eng. Sci., v. 58, p. 2035-2051.
Mathiesen, V., T. Solberg, and B. H. Hjertager, 2000, An experimental and computational study of multiphase flow behavior in a circulating fluidized bed: International Journal of Multiphase Flow, v. 26, p. 387-419.
McLaughlin, L. J., and M. J. Rhodes, 2001, Prediction of fluidized bed behavior in the presence of liquid bridges: Powder Technology, v. 114, p. 213-223.
Menon, N., and D. J. Durian, 1997, Particle motions in a gas-fluidized bed of sand: Phys. Rev. Lett., v. 79, p. 3407-3410.
Miller, A., and D. Gidaspow, 1992, Dense, vertical gas-solid flow in a pipe: AICHE J., v. 38, p. 1801-1815.
Montanero, J. M., V. Garzo, A. Santons, and J. J. Brey, 1999, Kinetic theory of simple granular shear flows of smooth hard spheres: J. Fluid Mech., v. 389, p. 391-411.
Moortel, T. V. D., E. Azario, R. Santini, and L. Tadrist, 1998, Experimental analysis of the gas-particle flow in a circulating fluidzied bed using a phase Doppler particle analyzer: Chem. Eng. Sci., v. 53, p. 1883-1899.
Moortel, T. V. D., R. Santini, L. Tadrist, and J. Pantaloni, 1997, Experimental study of the particle flow in a circulating fluidized bed using a phase Doppler particle

anayser: a new post-processing data algorithm: International Journal of Multiphase Flow, v. 23, p. 1189-1209.

Needham, D. J., and J. H. Merkin, 1983, The propagation of voidage disturbances in a uniform fluidized bed: J. Fluid Mech., v. 131, p. 427-454.

Needham, D. J., and J. H. Merkin, 1986, The existence and stability of quasi-steady periodic voidage waves in a fluidized bed: Z. Angew. Math. Phys., v. 37, p. 322-339.

Nieuwland, J. J., M. van Sint Annaland, J. A. M. Kuipers, and W. P. M. van Swaaij, 1996, Hydrodynamic modelling of gas/particle flows in riser reactors: AICHE J., v. 42, p. 1569-1582.

Nott, P. R., M. Alam, K. Agrawal, R. Jackson, and S. Sundaresan, 1999, The effect of boundaries on the plane Couette flow of granular materials: a bifurcation analysis: J. Fluid Mech., v. 397, p. 203-229.

Ocone, R., S. Sundaresan, and R. Jackson, 1993, Gas-particle flow in a duct of arbitrary inclination with particle-particle interactions: AICHE J., v. 39, p. 1261-1271.

Ogawa, S., A. Umemura, and N. Oshima, 1980, On the equations of fully fluidized granular materials: Z. Angew. Math. Phys., v. 31, p. 483-493.

Pigford, R. L., and T. Baron, 1965, Hydrodynamic stability of a fluidized bed: Ind. Eng. Chem. Fundam., v. 4, p. 81-87.

Pita, J. A., and S. Sundaresan, 1991, Gas-solid flow in vertical tubes: AICHE J., v. 37, p. 1009-1018.

Poilashenski, W., and J. J. C. Chen, 1997, Normal solid stress in fluidized beds: Powder Technology, v. 90, p. 13-23.

Rasuel, M. G., V. Rudolph, and M. Carsky, 1999, Segregation potential in binary gas fluidized beds: Powder Technology, v. 103, p. 175.

Rhode, M. J., and P. Laussmann, 1992, Characterising non-uniformities in gas-particle flow in the riser of a circulating fluidized bed: Powder Technology, v. 72, p. 277-284.

Richardson, J. F., and W. N. Zaki, 1954, Sedimentation and fluidization: Part I: Trans. Inst. Chem. Eng., v. 32, p. 35-53.

Richman, M. W., 1988, Boundary conditions based upon a modified Maxwellian velocity distribution for flows of identical, smooth, nearly elastic spheres: Acta Mehcanica, v. 75, p. 227-240.

Rudolph, M., T. Shinbrot, and R. Lueptow, 1998, A model of mixing and transport in Wavy-Taylor Couette flow: Physica D, v. 121, p. 163-174.

Samuelsberg, A., and Hjertager, 1996, An experimental and numerical study of flow patterns in a circulating fluidized bed reactor: International Journal of Multiphase Flow, v. 22, p. 575-591.

Sasa, S., and H. Hayakawa, 1992, Void-fraction dynamics in fluidization: Europhys. Lett., v. 17, p. 685-689.

Savage, S. B., and D. J. Jeffery, 1981, The stress tensor in a granular flow at high shear rates: J. Fluid Mech., v. 110, p. 255-272.

Savage, S. B., and C. K. K. Lun, 1988, Particle size segregation in clined chute flow of cohesionless granular solids: J. Fluid Mech., v. 189, p. 311-335.

Savage, S. B., and M. Sayed, 1984, Stresses developed by dry cohesionless granular materials sheared in annular shear cell: J. Fluid Mech., v. 142, p. 391-430.

Sela, N., I. Goldhirs, and S. H. Noskowicz, 1996, Kinetic theoretical study of a simply sheared two-dimensional granular gas to Burnett order: Phys. Fluids, v. 8, p. 2337-2353.
Sela, N., and I. Goldhirsh, 1998, Hydrodynamic equations for rapid flows of smooth inelastic spheres, to burnette order: J. Fluid Mech., v. 361, p. 41-74.
Sinclair, J. L., 1989, Vertical transport of gas and solids with radial solid density variations, PhD thesis, Princeton University.
Sinclair, J. L., and R. Jackson, 1989, Gas-particle flow in a vertical pipe with particle-particle interactions: AICHE J., v. 35, p. 1473-1486.
Srivastava, A., S. Agrawal, S. Sundaresan, S. B. Reddy Karri, and T. M. Knowlton, 1998, Dynamics of gas-particle flow in circulating fluidized beds: Powder Technology, v. 100, p. 173-182.
Stephen, L. C., Xue Liu and Benjamin J. Glasser, 2006, Instability induced clustering and segregation in high-shear granular Couette flows: Chem. Eng. Sci.
Tan, M. L., and I. Goldhirsch, 1997, Intercluster Interactions in rapid granular shear flows: Phys. Fluids, v. 9.
Tan, M. L., and I. Goldhirsch, 1998, Rapid granular glows as mesoscopic systems: Phys. Rev. Lett.. v. 81, p. 3022-3025.
Taylor, G. I., 1923, Stability of a viscous liquid contained between two rotating cylinders: Phil. Trans. Royal Soc. Lond., v. 223, p. 289-343.
Tsinontides, S., and R. Jackson, 1993, The mechanics of gas fluidized beds with an interval of stable fluidization: J. Fluid Mech., v. 255, p. 237-274.
Tsuji, Y., T. Kawaguchi, and T. Tanaka, 1993, Discrete particls simulation of two-dimensional fluidized bed: Powder Technology, v. 77, p. 79-87.
Wachem, B. G. M., J. C. Schouten, and C. M. Bleek, 2001, CFD modeling of gas-fluidized beds with a bimodal particle mixture: Particle Technology and Fluidization, v. 47, p. 1292-1302.
Wallis, G. B., 1991, The averaged Bernoulli equation and macroscopic equations of motion for the potential flow of a two-phase dispersion: International Journal of Multiphase Flow, v. 17, p. 683-695.
Wang, C. H., 1995, Instabilities in granular material flows, PH.D Thesis, Princeton University.
Wang, C. H., R. Jackson, and S. Sundaresan, 1996, Stability of bounded rapid shear flows of a granular material: J. Fluid Mech., v. 308, p. 31-62.
Wang, C. H., R. Jackson, and S. Sundaresan, 1997, Instabilities of fully developed rapid flow of a granular material in a channel: J. Fluid Mech., v. 342, p. 179-197.
Wang, C. H., and Z. Tong, 1998, Transient development of instabilities in bounded shear flow of granular materials: Chem. Eng. Sci., v. 53, p. 3803-3819.
Wang, C. H., and Z. Tong, 2001, On the density waves developed in gravity channel flows of granular materials: J. Fluid Mech., v. 435, p. 217-246.
Weinstein, H., M. Shao, and L. Wasserzug, 1984, Radial solid density variation in a fast fluidized bed: AICHE Symp. Ser., v. 80, p. 117-120.
Wen, C. Y., and Yu. H., 1966, A Generalized Method for Predicting the Minimum. Fluidization Velocity: AICHE J., v. 12, p. 610-612.
Wildman, R. D., and D. J. Parker, 2002, Coexistence of two granular temperatures in binary vibrofluidized beds: Phys. Rev. Lett., v. 88, p. 1-4.

Wilhelm, R. H., and M. Kwauk, 1948, Fluidization of solid particles: Chemical Engineering Progress, v. 44, p. 201-218.

Willits, J. T., and B. O. Arnarson, 1999, Kinetic theory of a binary mixture of nearly elastic disks: Phys. Fluids, v. 11, p. 3116-3122.

Wu, S. Y., and J. Baeyens, 1998, Segregation by size difference in gas fluidized beds: 98, v. Powder Technology, p. 139.

Xu, B. H., and A. B. Yu, 1997, Numerical simulation of the gas-solid flow in a fluidized bed by combining discrete particle method with computational fluid dynamics: Chem. Eng. Sci., v. 52, p. 2785-2809.

Yasuna, J. A., H. R. Moyer, S. Elliott, and J. L. Sinclair, 1995, Quantitative predictions of gas-particle flow in a vertical pipe with particle-particle interactions: Power Technology, v. 84, p. 23-34.

Yerushalmi, J., N. Cankurt, D. Geldart, and B. Liss, 1978, Flow regimes in verticla gas-solid contact system: AICHE Symp. Ser., v. 74, p. 1-13.

Zenit, R., M. L. Hunt, and C. E. Brennen, 1997, Collisional particle pressure measurements in solid-liquid flows: J. Fluid Mech., v. 353, p. 261-283.

Zhang, Y. H., and J. M. Reese, 2001, Particle-gas turbulence interactions in a kinetic theory approach to granular flows: International Journal of Multiphase Flow, v. 27, p. 1945-1964.

Zhou, H. S., G. Glamant, D. Gauthier, and J. D. Lu, 2002, Lagrangian approach for simulating the gas-particle flow structure in a circulating fluidized bed riser: International Journal of Multiphase Flow, v. 28, p. 1801-1821.

Appendix 1: Constitutive Relations in the Kinetic Theory Models

Constitutive relations in the Jenkins and Mancini (1989) model

$$\underline{\underline{P_{i-JM}}} = P_{Si}\underline{\underline{I}} - 2\mu_{i-JM}\underline{\underline{S_i}} \qquad \text{A.1.1}$$

$$P_{Si-JM} = T(n_i + \sum_{j=A,B} K_{ij}) \qquad \text{A.1.2}$$

$$P_{S-JM} = P_{SA-JM} + P_{SB-JM} \qquad \text{A.1.3}$$

$$K_{ij} = \frac{2\pi}{3} n_i n_j r_{ij}^3 g_{ij} \qquad \text{A.1.4}$$

$$\mu_{i-JM} = \frac{1}{2}(n_i T b_{i0} + \sum_{j=A,B} \mu_{ij}) \qquad \text{A.1.5}$$

$$\mu_{ij} = \frac{8}{15} n_i n_j \pi r_{ij}^3 g_{ij} [b_{i0} M_{ji} T + r_{ij} (\frac{2 m_i m_j T}{\pi m_{ij}})^{1/2}] \qquad \text{A.1.6}$$

$$b_{i0} = 5[b_i (n_i + \frac{2}{5} K_{ii} + \frac{4}{5} K_{ik} M_{ki}) + \frac{32}{3} (\pi M_{ki} M_{ik} / 2 / m_{ik})^{1/2} r_{ik}^2 (n_k + \frac{2}{5} K_{kk} + \frac{4}{5} K_{ik} M_{ik})][g_{ik} n_i n_k T^{1/2} (b_i b_k - \frac{512}{9} r_{ik}^4 \pi M_{ki} M_{ik} / m_{ik})]^{-1} \qquad \text{A.1.7}$$

$$b_i = 40 r_{ij}^2 (\frac{\pi M_{ik} M_{ki}}{2 m_{ik}})^{1/2} (\frac{2}{3} + \frac{2}{5}\frac{m_i}{m_k}) + 8 \frac{n_k g_{kk}}{n_i g_{ik}} r_{kk}^2 (\frac{\pi}{m_k})^{1/2} \qquad \text{A.1.8}$$

$$M_{ij} = m_i / m_{ij} \qquad m_{ij} = m_i + m_j \qquad \text{A.1.9}$$

$$g_{ij} = \frac{1}{1-\phi/\phi_{\max}} + \frac{3 r_i r_j}{r_i + r_j} \frac{\xi}{(1-\phi/\phi_{\max})^2} + 2(\frac{r_i r_j}{r_i + r_j})^2 \frac{\xi^2}{(1-\phi/\phi_{\max})^3} \qquad \text{A.1.10}$$

$$\phi = \pi(n_A r_A^2 + n_B r_B^2) \qquad \xi = \frac{4}{3}\pi(n_A r_A^2 + n_B r_B^2) \qquad \text{A.1.11}$$

$$\psi_i = \frac{\pi}{3} n_i n_k r_{ik}^3 g_{ik} (1 + e_p) T[(\frac{m_k - m_i}{m_{ik}}) \nabla \ln T + \nabla \ln \frac{n_i}{n_k}] \qquad \text{A.1.12}$$

$$\gamma = \sum_{i,j=A,B} 2g_{ij}r_{ij}^2 n_i n_j \frac{m_i}{m_{ij}}(1-e_p^2)(\frac{2\pi m_{ij}T^3}{m_i m_j})^{1/2} \quad \text{A.1.13}$$

$$\underline{q_{JM}} = -\kappa \nabla T \quad \text{A.1.14}$$

$$\kappa = -\frac{5}{4}\sum_{i=A,B}(n_i + \sum_{j=A,B}\frac{12\pi}{5}M_{ij}M_{ji}K_{ij})(\frac{2T}{m_i})^{1/2}a_{i1} + 2\sum_{i,j=A,B}K_{ij}r_{ij}(\frac{2m_i m_j T}{\pi m_{ij}^3})^{1/2} \quad \text{A.1.15}$$

$$a_{i1} = -\frac{15}{8\sqrt{\pi} n_i n_k g_{ik} r_{ik}^2 M_{ki}^{1/2}}(\frac{\alpha_i n_i K_i + 13M_{ik}M_{ki}n_k K_k}{\alpha_i \alpha_k - 169M_{ik}^2 M_{ki}^2}) \quad \text{A.1.16}$$

$$\alpha_i = 15M_{ki}^2 + 8M_{ik}M_{ki} + 6M_{ik}^2 + 2\sqrt{2}\frac{n_k g_{kk} r_{kk}^2}{n_i g_{ik} r_{ik}^2 M_{ik}^{1/2}} \quad \text{A.1.17}$$

$$K_i = 1 + \frac{2}{5}\pi n_i g_{ii} r_{ii}^3 + \frac{8\pi}{5}n_k r_{ik}^3 g_{ik} M_{ik} M_{ki} \quad \text{A.1.18}$$

$$\underline{\underline{S_i}} = \frac{1}{2}[\nabla \underline{v_i} + (\nabla \underline{v_i})^s - \frac{1}{3}\nabla \cdot \underline{v_i}\underline{\underline{I}} \quad \text{A.1.19}$$

We have assumed that $i \neq k$ in the above formulas.

Constitutive relations in the Iddir and Arastoopour (2005) model

$$\underline{\underline{P_i}} = P_{Si}\underline{\underline{I}} - 2\mu_i \underline{\underline{S_i}} \quad \text{A.1.20}$$

$$P_{Si} = n_i T_i + \sum_{j=A,B}\frac{\pi}{48}n_i n_j r_{ij}^3 g_{ij}(1+e_p)(\frac{m_i m_j}{m_{ij}})(\frac{m_i m_j}{T_i T_j})^{3/2} R_0 \quad \text{A.1.21}$$

$$P_S = P_{SA} + P_{SB} \quad \text{A.1.22}$$

$$\mu = \mu_A + \mu_B \quad \text{A.1.23}$$

$$\mu_i = \frac{1}{2}[\frac{5}{8r_{ii}^2 g_{ii}}(1 + \frac{4}{5}(1+e_p)g_{ii}\phi_i)^2(\frac{m_i T_i}{\pi})^{1/2} + \sum_{j=A,B}\mu_{ij}] \quad \text{A.1.24}$$

$$\mu_{ij} = \frac{\sqrt{\pi}}{120} n_i n_j r_{ij}^4 g_{ij} (1+e_p)(\frac{m_i m_j}{m_{ij}})^2 (\frac{m_i m_j}{T_i T_j})^{3/2} (\frac{1}{T_i} + \frac{1}{T_j}) R_1 \qquad \text{A.1.25}$$

$$\underline{F_{D1i}} = \sum_{j=A,B} \underline{F_{Dij}} \qquad \text{A.1.26}$$

$$\underline{F_{D1ij}} = \frac{r_{ij}^2}{4} n_i n_j g_{ij} (1+e_p)(\frac{m_i m_j}{m_{ij}})(\frac{m_i m_j}{T_i T_j})^{3/2} \{\frac{r_{ij}\pi}{12} R_0 (\nabla \ln \frac{n_i}{n_j} - \frac{3}{2} \nabla \ln \frac{T_i}{T_j})$$
$$+ \frac{r_{ij}\pi}{16} [(\frac{m_j \nabla T_j}{T_j^2} - \frac{m_i \nabla T_i}{T_i^2}) R_{10} + \frac{5 m_i m_j}{192 m_{ij}^2} (\frac{m_i \nabla T_j}{T_j^2} - \frac{m_j \nabla T_i}{T_i^2}) R_3 + \qquad \text{A.1.27}$$
$$\frac{5 m_i m_j}{96 m_{ij}} B_{ij} (\frac{\nabla T_j}{T_j^2} + \frac{\nabla T_i}{T_i^2}) R_4]\}$$

$$\underline{F_{D2i}} = \frac{r_{ij}^2}{4} n_i n_k g_{ik} (1+e_p)(\frac{m_i m_k}{m_{ik}})(\frac{m_i m_k}{T_i T_k})^{3/2} [\pi R_2 (\underline{v_k} - \underline{v_i})] \qquad \text{A.1.28}$$

$$N_i = \sum_{j=A,B} N_{ij} \qquad \text{A.1.29}$$

$$N_{ij} = \frac{3 r_{ij}^2}{4} n_i n_j g_{ij} (1+e_p)(\frac{m_i m_j}{m_{ij}})(\frac{m_i m_j}{T_i T_j})^{3/2} [B_{ij} \sqrt{\pi} R_5 + (e_p - 1) \frac{m_j}{m_{ij}} \frac{\sqrt{\pi}}{6} R_1] \qquad \text{A.1.30}$$

$$q = q_A + q_B \qquad \text{A.1.31}$$

$$\underline{q_{1i}} = \frac{-75}{64 r_{ii}^2 g_{ii}} (1 + \frac{3}{5}(1+e_p)^2 g_{ii} \phi_i)^2 (\frac{T_i}{m_i \pi})^{1/2} \nabla T_i + \sum_{j=A,B} \underline{q_{1cij}} \qquad \text{A.1.32}$$

$$\underline{q_{1cij}} = \frac{r_{ij}^3}{2} n_i n_j g_{ij} (1+e_p)(\frac{m_i m_j}{m_{ij}})(\frac{m_i m_j}{T_i T_j})^{3/2} \{\frac{r_{ij}\sqrt{\pi}}{16} [R_5 (\nabla \ln \frac{n_i}{n_j} + \frac{3}{2} \nabla \ln \frac{T_i}{T_j}) B_{ij}$$
$$+ \frac{B_{ij}}{8} [(\frac{m_j \nabla T_j}{T_j^2} - \frac{m_i \nabla T_i}{T_i^2}) R_6 + B_{ij} \frac{m_i m_j}{m_{ij}^2} (\frac{m_i \nabla T_j}{T_j^2} - \frac{m_j \nabla T_i}{T_i^2}) R_7 +$$
$$\frac{m_i m_j}{2 m_{ij}} (\frac{\nabla T_j}{T_j^2} - \frac{\nabla T_i}{T_i^2}) R_9] + \frac{m_j}{8 m_{ij}} (1-e_p)[\frac{r_{ij}\pi}{6} (\nabla \ln \frac{n_i}{n_j} + \frac{3}{2} \nabla \ln \frac{T_i}{T_j}) R_1 \qquad \text{A.1.33}$$
$$+ \frac{r_{ij}\sqrt{\pi}}{8} [(\frac{m_j \nabla T_j}{T_j^2} - \frac{m_i \nabla T_i}{T_i^2}) R_9 + \frac{r_{ij}\sqrt{\pi}}{2} \frac{m_i^2}{m_{ij}^2} (\frac{m_i \nabla T_j}{T_j^2} - \frac{m_j \nabla T_i}{T_i^2}) R_8 +$$
$$r_{ij} \sqrt{\pi} B_{ij} \frac{m_i m_j}{m_{ij}} (\frac{\nabla T_j}{T_j^2} + \frac{\nabla T_i}{T_i^2}) R_7]\}$$

$$\underline{q_{2i}} = \frac{r_{ik}^3}{2} n_i n_k g_{ik} (1+e_p)(\frac{m_i m_k}{m_{ik}})(\frac{m_i m_k}{T_i T_k})^{3/2} [\frac{m_k}{8m_{ik}}(1-e_p)\frac{3\pi}{10}(v_i - v_k) R_0] \qquad \text{A.1.34}$$

$$A_{ij} = \frac{m_i T_j + m_j T_i}{2T_i T_j} \qquad B_{ij} = \frac{m_i m_j (T_j - T_i)}{2m_{ij} T_i T_j}$$

$$D_{ij} = \frac{m_i m_j (m_j T_j + m_i T_i)}{2m_{ij\,i}^2 T_j} \qquad \text{A.1.35}$$

$$R_0 = \frac{1}{A_{ij}^{3/2} D_{ij}^{5/2}} + \frac{15B_{ij}^2}{2A_{ij}^{5/2} D_{ij}^{7/2}} + \frac{175B_{ij}^4}{8A_{ij}^{7/2} D_{ij}^{9/2}} + \cdots \qquad \text{A.1.36}$$

$$R_1 = \frac{1}{A_{ij}^{3/2} D_{ij}^{3}} + \frac{9B_{ij}^2}{2A_{ij}^{5/2} D_{ij}^{4}} + \frac{30B_{ij}^4}{2A_{ij}^{7/2} D_{ij}^{5}} + \cdots \qquad \text{A.1.37}$$

$$R_2 = \frac{1}{2A_{ij}^{3/2} D_{ij}^{2}} + \frac{3B_{ij}^2}{A_{ij}^{5/2} D_{ij}^{3}} + \frac{15B_{ij}^4}{2A_{ij}^{7/2} D_{ij}^{4}} + \cdots \qquad \text{A.1.38}$$

$$R_3 = \frac{1}{A_{ij}^{3/2} D_{ij}^{7/2}} + \frac{21B_{ij}^2}{2A_{ij}^{5/2} D_{ij}^{9/2}} + \frac{315B_{ij}^4}{8A_{ij}^{7/2} D_{ij}^{11/2}} + \cdots \qquad \text{A.1.39}$$

$$R_4 = \frac{3}{A_{ij}^{5/2} D_{ij}^{7/2}} + \frac{35B_{ij}^2}{2A_{ij}^{7/2} D_{ij}^{9/2}} + \frac{441B_{ij}^4}{8A_{ij}^{9/2} D_{ij}^{11/2}} + \cdots \qquad \text{A.1.40}$$

$$R_5 = \frac{1}{A_{ij}^{5/2} D_{ij}^{3}} + \frac{5B_{ij}^2}{A_{ij}^{7/2} D_{ij}^{4}} + \frac{14B_{ij}^4}{A_{ij}^{9/2} D_{ij}^{5}} + \cdots \qquad \text{A.1.41}$$

$$R_6 = \frac{1}{A_{ij}^{7/2} D_{ij}^{3}} + \frac{7B_{ij}^2}{A_{ij}^{9/2} D_{ij}^{4}} + \frac{126B_{ij}^4}{5A_{ij}^{11/2} D_{ij}^{5}} + \cdots \qquad \text{A.1.42}$$

$$R_7 = \frac{3}{2A_{ij}^{5/2} D_{ij}^{4}} + \frac{10B_{ij}^2}{A_{ij}^{7/2} D_{ij}^{5}} + \frac{35B_{ij}^4}{A_{ij}^{9/2} D_{ij}^{6}} + \cdots \qquad \text{A.1.43}$$

$$R_8 = \frac{1}{2A_{ij}^{3/2} D_{ij}^{4}} + \frac{6B_{ij}^2}{A_{ij}^{5/2} D_{ij}^{5}} + \frac{25B_{ij}^4}{A_{ij}^{7/2} D_{ij}^{6}} + \cdots \qquad \text{A.1.44}$$

$$R_9 = \frac{1}{A_{ij}^{5/2} D_{ij}^3} + \frac{15B_{ij}^2}{A_{ij}^{7/2} D_{ij}^4} + \frac{70B_{ij}^4}{A_{ij}^{9/2} D_{ij}^5} + \cdots \quad \text{A.1.45}$$

$$R_{10} = \frac{1}{A_{ij}^{5/2} D_{ij}^{5/2}} + \frac{25B_{ij}^2}{2A_{ij}^{7/2} D_{ij}^{7/2}} + \frac{1225B_{ij}^4}{24A_{ij}^{9/2} D_{ij}^{9/2}} + \cdots \quad \text{A.1.46}$$

$$C_n = \frac{r_{AB}^2}{4} n_A n_B g_{AB} (1+e_p)(\frac{m_A m_B}{m_{AB}})(\frac{m_A m_B}{T_A T_B})^{3/2} (\frac{r_{AB}\pi}{12} R_0) \quad \text{A.1.47}$$

$$C_{TA} = \frac{r_{AB}^2}{4} n_A n_B g_{AB} (1+e_p)(\frac{m_A m_B}{m_{AB}})(\frac{m_A m_B}{T_A T_B})^{3/2} [(n_B m_B + n_A m_A)(-\frac{3}{2}\frac{r_{AB}\pi}{12} R_0 \frac{1}{T_A} - \frac{r_{AB}\pi}{16} R_{10} \frac{m_A}{T_A^2} - \frac{5m_A m_B}{192m_{AB}^2} R_3 \frac{m_B}{T_A^2} + \frac{5m_A m_B}{96m_{AB}} R_4 B_{AB} \frac{\nabla T_A}{T_A^2})] \quad \text{A.1.48}$$

$$C_{TB} = \frac{r_{AB}^2}{4} n_A n_B g_{AB} (1+e_p)(\frac{m_A m_B}{m_{AB}})(\frac{m_A m_B}{T_A T_B})^{3/2} [(n_B m_B + n_A m_A)(-\frac{3}{2}\frac{r_{AB}\pi}{12} R_0 \frac{1}{T_B} - \frac{r_{AB}\pi}{16} R_{10} \frac{m_B}{T_B^2} - \frac{5m_A m_B}{192m_{AB}^2} R_3 \frac{m_A}{T_B^2} + \frac{5m_A m_B}{96m_{AB}} R_4 B_{AB} \frac{\nabla T_B}{T_B^2})] \quad \text{A.1.49}$$

We have assumed that $i \neq k$ in the above formulas.

Appendix 2: Orthogonal Collocation Method

Collocation method is based on the concept of interpolation of unequally spaced points (Constantinides, 1987). A function, usually a polynomial, is used to approximate the solution of a differential equation in the range of integration. Then the coefficients of this function are determined based on a set of base points.

The orthogonal collocation method, which is an extension of the collocation method, provides a mechanism for automatically picking the collocation points by making use of orthogonal polynomials. Usually for a second-order two-point boundary-value problem, *N*+2 collocation points are required, including two known boundary points and *N* internal points that are determined from the zero roots of a orthogonal polynomial. Then the function *y(x)* can be written based on this *N*+2 points,

$$y(x_j) = \sum_{i=1}^{n+2} d_i x_j^{i-1} \qquad \text{A.2.1}$$

and the derivatives of *y* are taken as

$$\frac{dy(x_j)}{dx} = \sum_{i=1}^{n+2} d_i (i-1) x_j^{i-2} \qquad \frac{d^2 y(x_j)}{dx^2} = \sum_{i=1}^{n+2} d_i (i-1)(i-2) x_j^{i-3} \qquad \text{A.2.2}$$

These equations can be written in the matrix form

$$\frac{dy}{dx} = CE^{-1}y = Py \qquad \frac{d^2 y}{dx^2} = DE^{-1}y = Qy \qquad \text{A.2.3}$$

where

$$E_{ji} = x_j^{i-1} \qquad C_{ji} = (i-1)x_j^{i-2} \qquad D_{ji} = (i-1)(i-2)x_j^{i-3} \qquad \text{A.2.4}$$

To get the solutions for the steady, fully developed gas-particle flows in a vertical duct (see Chapter 3), a numerical method based on the orthogonal collocation approach is set up. The internal collocation points are chosen from the roots of the Legendre polynomials. We use Matlab 6.0 software to get the coefficient matrix *P* and *Q* because a large calculation error occurs in the Fortran code when the number of collocation points is greater than 13.

The governing equations have a form of a nonlinear set of coupled ordinary differential equations (see Equation 3.1-3.4) that are solved as a two-point boundary value problem (see Equation 3.5-3.8) with the specification of values for dimensionless parameters that are determined by the gas-solid system under the consideration. Once the gas-particle system is specified, solutions to the equations depend on two remaining parameters, the dimensionless pressure gradient dp^*/dx^* and the dimensionless integration constant C_1. Solution profiles for the gas velocity, the solid velocity, the solids fraction, and the granular temperature can be generated for the range of operating conditions, given values for dp^*/dx^* and C_1.

An ad-hoc linearization method on the governing equations and boundary conditions is used to solve the above nonlinear system. For example, the second term in the pseudo-thermal energy balance (see Equation 3.4) is approximated to be

$$D^2 f_5 (T^*)^{\frac{3}{2}} = [D^2 f_5 (T^*)^{\frac{1}{2}}] T^*$$

and the last term in the gas phase boundary condition (see Equation 3.7) is approximated to be

$$2G \frac{\phi_{\max}}{\phi} T^* u_x^* = [G \frac{\phi_{\max}}{\phi} T^*] u_x^* + [G \frac{\phi_{\max}}{\phi} u_x^*] T^*$$

where the terms in brackets are evaluated with approximate values obtained from the previous step. The trial profile for the solids fraction is generated from the granular temperature profile, using the algebraic Equation 3.9. Firstly this equation is solved by the bisection method. Then the Netwon-Raphson method is used to get more accurate solutions.

The linearized system of equations is solved by the orthogonal collocation method. After substituting the first and second order derivatives of u_x^*, v_x^*, T^*, and ϕ^* based on the matrix P and Q, the collocation equations to be solved for the point j take the form,

$$-[\frac{BDE}{C}+\frac{BD}{C}\frac{dp^*}{dx^*}]-[\frac{BD(1-E)}{C}f_0](u^*_{x(j)}-v^*_{x(j)})+[\mu^*_g](\sum_{i=1}^{N+2}Q_{ji}u^*_{x(i)})=0 \qquad \text{A.2.5}$$

$$[\frac{D(1-E)}{C}f_0](u^*_{x(j)}-v^*_{x(j)})-[\frac{D(1-E)}{C}f_0\phi_j]+[\frac{1}{2}f_3(T^*_j)^{\frac{1}{2}}](\sum_{i=1}^{N+2}Q_{ji}v^*_{x(i)})+[\frac{1}{8}f_3(T^*_j)^{-\frac{1}{2}}$$
$$+\sum_{i=1}^{N+2}P_{ji}T^*_i\frac{1}{2}(T^*_j)^{\frac{1}{2}}\frac{df_3}{d\phi}\sum_{i=1}^{N+2}P_{ji}\phi_i](\sum_{i=1}^{N+2}P_{ji}v^*_{x(i)})+[\frac{1}{8}f_3(T^*_j)^{-\frac{1}{2}}\sum_{i=1}^{N+2}P_{ji}v^*_{x(i)}](\sum_{i=1}^{N+2}P_{ji}T^*_i)=0$$

A.2.6

$$[f_4(T^*_j)^{\frac{1}{2}}](\sum_{i=1}^{N+2}Q_{ji}T^*_i)+[(T^*_j)^{\frac{1}{2}}\frac{df_4}{d\phi}\sum_{i=1}^{N+2}P_{ji}\phi_i+\frac{1}{2}f_4(T^*_j)^{-\frac{1}{2}}\sum_{i=1}^{N+2}P_{ji}T^*_i+\frac{1}{2}f_{4h}(T^*_j)^{\frac{1}{2}}{}_i]$$
$$\sum_{i=1}^{N+2}P_{ji}\phi(\sum_{i=1}^{N+2}P_{ji}T^*_i)-[D^2f_5(T^*_j)^{\frac{1}{2}}+\frac{(1-E)D}{C}3f_{01}-f_{4h}(T^*_j)^{\frac{1}{2}}\sum_{i=1}^{N+2}Q_{ji}\phi_i-(T^*_j)^{\frac{1}{2}}\frac{df_{4h}}{d\varepsilon_s}$$
$$(\sum_{i=1}^{N+2}P_{ji}\phi_i)^2]T^*_j+[\frac{D^2}{B^2}f_{02}(T^*_j)^{-\frac{1}{2}}(u^*_{x(j)}-v^*_{x(j)})](u^*_{x(j)}-v^*_{x(j)})+[\frac{1}{2}f_3(T^*_j)^{\frac{1}{2}}\sum_{i=1}^{N+2}P_{ji}v^*_{x(i)}]$$
$$(\sum_{i=1}^{N+2}P_{ji}v^*_{x(i)})=0$$

A.2.7

where j is from 2 to N+1. Note that there are 3N equations provided at the N interior collocation points and 3(N+2) unknowns. Therefore, to solve the system we need another 6 equations, which are provided at the boundary points. At $y^*=0$, the symmetric boundary conditions take the form,

$$\sum_{i=1}^{N+2} P_{1i}u^*_{x(i)} = 0 \qquad \sum_{i=1}^{N+2} P_{1i}v^*_{x(i)} = 0 \qquad \sum_{i=1}^{N+2} P_{1i}T^*_i = 0 \qquad \text{A.2.8}$$

At $y^*=1$, the boundary conditions for the solid phase momentum balance, the pseudo-thermal energy balance and the gas phase momentum balance have the form,

$$\sum_{i=1}^{N+2} P_{Ni}v^*_{x(i)} + [D\varphi' f_6]v^*_{x(N+2)} = 0 \qquad \text{A.2.9}$$

$$\sum_{i=1}^{N+2} P_{Ni}T^*_i - [D\varphi' f_7 v^*_{x(N+2)}]v^*_{x(N+2)} + [D(1-e_w^2)f_8 + \frac{f_{4h}}{f_4}\sum_{i=1}^{N+2} P_{Ni}\phi_i]T^*_{N+2} = 0 \qquad \text{A.2.10}$$

$$[\mu^*_{eg}]\sum_{i=1}^{N+2} P_{Ni}v^*_{x(i)} + [\frac{BDG(1-E)}{C}\frac{\phi_{N+2}}{\phi_{\max}} f_0](u^*_{x(N+2)} - v^*_{x(N+2)}) + [\frac{BDG}{C}\frac{\phi_{N+2}}{\phi_{\max}}\frac{dp^*_f}{dy^*}] + [G\frac{\phi_{\max}}{\phi_{N+2}}T^*_{N+2}]u^*_{x(N+2)} + [G\frac{\phi_{\max}}{\phi_{N+2}}u^*_{x(N+2)}]T^*_{N+2} = 0 \qquad \text{A.2.11}$$

Equations 2.5-2.11 are solved to yield new values for the variable vectors u^*_x, v^*_x, T^*. Then the algebraic equation is solved to get new values for ϕ^* based on the new values of the granular temperature. This procedure is iterated until changes in the computed values of u^*_x, v^*_x, T^* are less than 10^{-5} between 5 successive iterations. The accuracy of the numerical scheme is ascertained by increasing the number of collocation points until the solutions are independent with the variation of the number of collocation points. It is found that the larger the bed width, the larger the number of collocation points required to achieve convergence.

To ensure accurate computation of the engienvalues, our simulation code is checked by favorable comparison with published results of Wang C.H (1995), who solved steady state solutions for Couette granular flow with a monodisperse particle mixture. The governing equations have the form:

$$\frac{d}{dy^*}\left(\frac{1}{2} f_3 (T^*)^{\frac{1}{2}} \frac{dv_x^*}{dy^*}\right) = 0 \qquad \text{A.2.12}$$

$$\frac{d}{dy^*}(f_1 T^*) = 0 \qquad \text{A.2.13}$$

$$\frac{d}{dy^*}\left(f_4 (T^*)^{\frac{1}{2}} \frac{dT^*}{dy^*}\right) - D^2 f_5 (T^*)^{\frac{3}{2}} + f_3 (T^*)^{\frac{1}{2}} \frac{1}{2}\left(\frac{dv_x^*}{dy^*}\right)^2 = 0 \qquad \text{A.2.14}$$

The boundary conditions take the form:

$$\frac{dv_x^*}{dy^*} + D\varphi' f_6 (v_x^* - 0.5) = 0 \qquad \text{A.2.15}$$

$$\frac{dT^*}{dy^*} - D\varphi' f_7 (v_x^* - 0.5)^2 + D(1 - e_w^2) f_8 T^* = 0 \qquad \text{A.2.16}$$

The dimensionless functions $f_1 - f_8$ can be found in Table 3.1.

The comparison of the results from Wang's work (1995) and our simulations is shown in Fig A2.1. Clearly, our simulation results match Wang's data quite well. Another numerical scheme, namely, a second-order finite difference method is also used to ascertain the correctness of the solutions. From Fig A2.2, it can be seen that for granular Couette flows the collocation method and the finite difference approach generate the same flow profiles.

Figure A2.1: Comparison of Wang's simulation results and our simulation results for solids fraction profiles in a granular Couette flow (e_p=1, $\bar{\phi}$=0.35). Curves represent our simulation results; symbols represent Wang's simulation results.

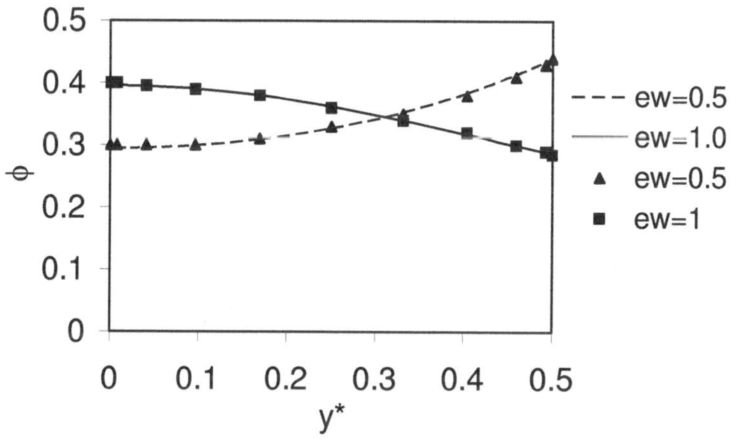

Figure A2.2: Comparison of simulation results from the collocation approach and finite difference method for granular Couette flow (e_p=1, e_w=1, $\bar{\phi}$=0.35). a) solids fraction profile b) granular temperature profile c) gas velocity profile

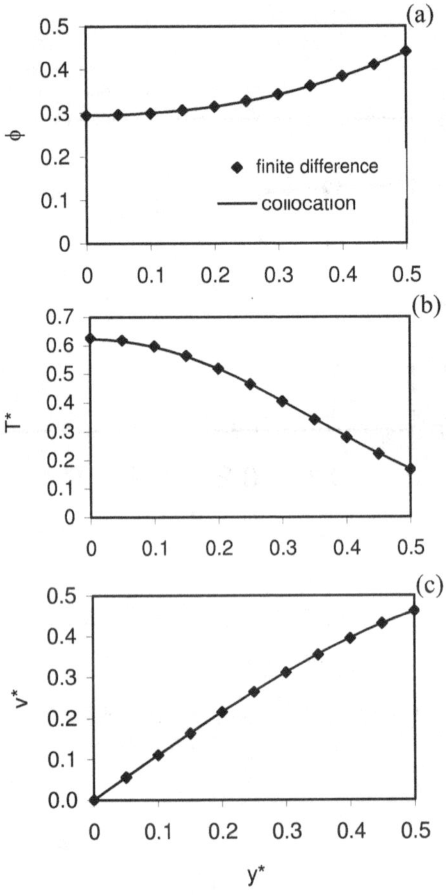

Appendix 3: Linear Stability Equations for Bounded Gas-Particle Flows

Model Equations

$$A1\hat{\phi} + A2\hat{u}_x + (A4 + A3\frac{d}{dy})\hat{u}_y = s\hat{\phi} \quad \text{A.3.1}$$

$$B1\hat{\phi} + B2\hat{v}_x + (B4 + B3\frac{d}{dy})\hat{v}_y = -s\hat{\phi} \quad \text{A.3.2}$$

$$D9\hat{\phi} + (D7 + D1\frac{d}{dy})\hat{u}_x + (D3 + D8\frac{d}{dy} + D4\frac{d^2}{dy^2})\hat{u}_y + D5\hat{v}_y + D6\hat{P}_g = -D2s\hat{u}_y \quad \text{A.3.3}$$

$$(C7 + C10\frac{d}{dy})\hat{\phi} + (C2 + C3\frac{d^2}{dy^2} + C9\frac{d}{dy})\hat{u}_x + (C4 + C5\frac{d}{dy})\hat{u}_y + C6\hat{v}_x + (C8\frac{d}{dy})\hat{P}_g = -C1s\hat{u}_x \quad \text{A.3.4}$$

$$(E1 + E2\frac{d}{dy})\hat{\phi} + E4\hat{u}_x + E5\hat{u}_y + (E7 + E8\frac{d}{dy} + E9\frac{d^2}{dy^2})\hat{v}_x + (E10 + E11\frac{d}{dy})\hat{v}_y + (E12 + E13\frac{d}{dy})\hat{T} = -E3s\hat{u}_x - E6s\hat{v}_x \quad \text{A.3.5}$$

$$(G1 + G2\frac{d}{dy})\hat{\phi} + +G4\hat{u}_y + (G9 + G10\frac{d}{dy})\hat{v}_x + (G6 + G7\frac{d}{dy} + G8\frac{d^2}{dy^2})\hat{v}_y + (G11 + G12\frac{d}{dy})\hat{T} = -G3s\hat{v}_x - G5s\hat{v}_y \quad \text{A.3.6}$$

$$(H1 + H2\frac{d}{dy})\hat{\phi} + +H4\hat{u}_x + (H5 + H6\frac{d}{dy})\hat{v}_x + (H7 + H8\frac{d}{dy})\hat{v}_y + (H10 + H11\frac{d}{dy} + H12\frac{d^2}{dy^2})\hat{T} = -H9s\hat{T} \quad \text{A.3.7}$$

Boundary Conditions

(at $y = 1$)

$$P1\hat{\phi}+(P2+\frac{d}{dy})\hat{v}_x=0 \qquad \text{A.3.8}$$

$$\hat{v}_y=0 \qquad \text{A.3.9}$$

$$Q1\hat{\phi}+Q3\hat{v}_x+(Q4+\frac{d}{dy})\hat{T}=0 \qquad \text{A.3.10}$$

$$R1\hat{\phi}+(R2+R3\frac{d}{dy})\hat{u}_x+R4\hat{v}_x+R5\hat{T}+R6\hat{P}_g=0 \qquad \text{A.3.11}$$

$$\hat{v}_y=0 \qquad \text{A.3.12}$$

(at *y=-1*)

$$P1\hat{\phi}+(P2-\frac{d}{dy})\hat{v}_x=0 \qquad \text{A.3.13}$$

$$\hat{v}_y=0 \qquad \text{A.3.14}$$

$$Q1\hat{\phi}+Q3\hat{v}_x+(Q4-\frac{d}{dy})\hat{T}=0 \qquad \text{A.3.15}$$

$$R1\hat{\phi}+(R2-R3\frac{d}{dy})\hat{u}_x+R4\hat{v}_x+R5\hat{T}+R6\hat{P}_g=0 \qquad \text{A.3.16}$$

$$\hat{v}_y=0 \qquad \text{A.3.17}$$

Coefficients in the Equations

$$A1=-ik_x u_x^0$$

$$A2=ik_x(1-\phi^0)$$

$$A3=1-\phi^0$$

$$A4 = -\frac{d\phi^0}{dy}$$

$$B1 = ik_x v_x^0$$

$$B2 = ik_x \phi^0$$

$$B3 = \phi^0$$

$$B4 = \frac{d\phi^0}{dy}$$

$$C1 = BED$$

$$C2 = BEDik_x u_x^0 + \frac{BD(1-E)}{C} f_0(\phi^0) + \frac{4}{3}\mu_{eg}(\phi^0) k_x^2$$

$$C3 = -\mu_{eg}(\phi^0)$$

$$C4 = BED\frac{du_x^0}{dy} - ik_x \frac{d\mu_{eg}(\phi^0)}{d\phi}\frac{d\phi^0}{dy}$$

$$C5 = -\frac{1}{3} ik_x \mu_{eg}(\phi^0)$$

$$C6 = -\frac{BD(1-E)}{C} f_0(\phi^0)$$

$$C7 = \frac{BD(1-E)}{C}\frac{df_0(\phi^0)}{d\phi}(u_x^0 - v_x^0) - \frac{d\mu_{eg}(\phi^0)}{d\phi}\frac{d^2 u_x^0}{dy^2} - \frac{d^2\mu_{eg}(\phi^0)}{d\phi^2}\frac{d\phi^0}{dy}\frac{du_x^0}{dy}$$

$$C8 = \frac{BD}{C} ik_x$$

$$C9 = -\frac{d\mu_{eg}(\phi^0)}{d\phi}\frac{d\phi^0}{dy}$$

$$C10 = -\frac{d\mu_{eg}(\phi^0)}{d\phi}\frac{du_x^0}{dy}$$

$$D1 = -\frac{1}{3} ik_x \mu_{eg}(\phi^0)$$

$$D2 = BED$$

$$D3 = ik_x BEDu_x^0 + k_x^2 \mu_{eg}(\phi^0) + \frac{BD(1-E)}{C} f_0(\phi^0)$$

$$D4 = -\frac{4}{3} \mu_{eg}(\phi^0)$$

$$D5 = -\frac{BD(1-E)}{C} f_0(\phi^0)$$

$$D6 = \frac{BD}{C}$$

$$D7 = \frac{2}{3} ik_x \frac{d\mu_{eg}(\phi^0)}{d\phi} \frac{d\phi^0}{dy}$$

$$D8 = -\frac{4}{3} \frac{d\mu_{eg}(\phi^0)}{d\phi} \frac{d\phi^0}{dy}$$

$$D9 = -ik_x \frac{d\mu_{eg}(\phi^0)}{d\phi} \frac{du_x^0}{dy}$$

$$E1 = -\frac{D(1-E)}{C} \frac{df_0(\phi^0)}{d\phi} (u_x^0 - v_x^0) + D\frac{df_1(\phi^0)}{d\phi} T^0 ik_x - \frac{1}{2} \frac{dv_x^0}{dy} (T^0)^{\frac{1}{2}} \frac{d^2 f_3(\phi^0)}{d\phi^2} \frac{d\phi}{dy}$$
$$-\frac{1}{2} \frac{d^2 v_x^0}{dy^2} (T^0)^{\frac{1}{2}} \frac{df_3(\phi^0)}{d\phi} - \frac{1}{4} \frac{dv_x^0}{dy} (T^0)^{-\frac{1}{2}} \frac{df_3(\phi^0)}{d\phi} \frac{dT^0}{dy} + \frac{D(1-E)}{C}$$

$$E2 = -\frac{1}{2} \frac{dv_x^0}{dy} (T^0)^{\frac{1}{2}} \frac{df_3(\phi^0)}{d\phi}$$

$$E3 = -DE\phi^0$$

$$E4 = -DE\phi^0 ik_x u_x^0 - \frac{D(1-E)}{C} f_0(\phi^0)$$

$$E5 = -DE\phi^0 \frac{du_x^0}{dy}$$

$$E6 = D\phi^0$$

$$E7 = D\phi^0 ik_x v_x^0 + \frac{D(1-E)}{C} f_0(\phi^0) + f_2(\phi^0)(T^0)^{\frac{1}{2}} k_x^2 + \frac{2}{3} f_3(\phi^0)(T^0)^{\frac{1}{2}} k_x^2$$

$$E8 = -\frac{1}{2}(\frac{df_3(\phi^0)}{dy}(T^0)^{\frac{1}{2}} + \frac{1}{2} f_3(\phi^0)(T^0)^{-\frac{1}{2}} \frac{dT^0}{dy})$$

$$E9 = -\frac{1}{2} f_3(\phi^0)(T^0)^{\frac{1}{2}}$$

$$E10 = D\phi^0 \frac{du_x^0}{dy} - \frac{1}{2}\frac{d}{dy}(f_3(\phi^0)(T^0)^{\frac{1}{2}})ik_x$$

$$E11 = -f_2(\phi^0)(T^0)^{\frac{1}{2}} ik_x - \frac{1}{6} f_3(\phi^0)(T^0)^{\frac{1}{2}} ik_x$$

$$E12 = Df_1(\phi^0)ik_x - \frac{d}{dy}(\frac{1}{4}(T^0)^{-\frac{1}{2}} f_3(\phi^0)\frac{dv_x^0}{dy})$$

$$E13 = -\frac{1}{4}(T^0)^{-\frac{1}{2}} f_3(\phi^0)\frac{dv_x^0}{dy}$$

$$G1 = D\frac{d}{dy}(T^0 \frac{df_1(\phi^0)}{d\phi}) - \frac{1}{2}\frac{df_3(\phi^0)}{d\phi}(T^0)^{\frac{1}{2}} \frac{dv_x^0}{dy} ik_x$$

$$G2 = DT^0 \frac{df_1(\phi^0)}{d\phi}$$

$$G3 = -DE\phi^0$$

$$G4 = -DE\phi^0 u_x^0 ik_x - \frac{D(1-E)}{C} f_0(\phi^0)$$

$$G5 = D\phi^0$$

$$G6 = D\phi^0 v_x^0 ik_x + \frac{D(1-E)}{C} f_0(\phi^0) + f_0(\phi^0)(T^0)^{\frac{1}{2}} ik_x$$

$$G7 = -\frac{d}{dy}(f_2(\phi^0)(T^0)^{\frac{1}{2}}) - \frac{d}{dy}(\frac{2}{3} f_3(\phi^0)(T^0)^{\frac{1}{2}})$$

$$G8 = -f_2(\phi^0)(T^0)^{\frac{1}{2}} - \frac{2}{3} f_3(\phi^0)(T^0)^{\frac{1}{2}}$$

$$H1 = \frac{d}{dy}(\frac{df_4(\phi^0)}{d\phi}(T^0)^{\frac{1}{2}} \frac{dT^0}{dy}) - D^2 \frac{df_5(\phi^0)}{d\phi}(T^0)^{\frac{3}{2}} - 3\frac{D(1-E)}{C}\frac{df_0(\phi^0)}{d\phi}(T^0)^{\frac{1}{2}}$$
$$+ \frac{D^2}{B^2}\frac{df_{02}(\phi^0)}{d\phi}(T^0)^{-\frac{1}{2}}(u_x^0 - v_x^0)^2 + \frac{1}{2}(\frac{dv_x^0}{dy})^2 \frac{df_3(\phi^0)}{d\phi} T^0$$

$$H2 = \frac{df_4(\phi^0)}{d\phi}(T^0)^{\frac{1}{2}} \frac{dT^0}{dy}$$

$$H4 = \frac{D^2}{B^2}\frac{df_{02}(\phi^0)}{d\phi}(T^0)^{-\frac{1}{2}} 2(u_x^0 - v_x^0)$$

$$H5 = -H4 - Df_1(\phi^0) T^0 ik_x$$

$$H6 = f_3(\phi^0)(T^0)^{\frac{1}{2}} \frac{dv_x^0}{dy}$$

$$H7 = -\frac{3}{2} D \frac{dT^0}{dy} \phi^0 + H6ik_x$$

$$H8 = -Df_1(\phi^0) T^0$$

$$H9 = -\frac{3}{2} D\phi^0$$

$$H10 = -\frac{3}{2} D\phi^0 v_x^0 ik_x - f_4(\phi^0)(T^0)^{\frac{1}{2}} k_x^2 + \frac{d}{dy}(f_4(\phi^0)\frac{1}{2}(T^0)^{-\frac{1}{2}}\frac{dT^0}{dy}) - \frac{3}{2} D^2 f_5(\phi^0)(T^0)^{\frac{1}{2}} -$$
$$3\frac{D(1-E)}{C} f_{01}(\phi^0) + \frac{D^2}{B^2} f_{02}(\phi^0)(-\frac{1}{2})(T^0)^{-\frac{3}{2}}(u_x^0 - v_x^0)^2 + \frac{1}{4}(\frac{dv_x^0}{dy})^2 f_3(\phi^0)(T^0)^{-\frac{1}{2}}$$

$$H11 = \frac{d}{dy}(f_4(\phi^0)(T^0)^{\frac{1}{2}})$$

$$H12 = f_4(\phi^0)(T^0)^{\frac{1}{2}}$$

$$P1 = D\varphi' \frac{df_6(\phi^0)}{d\phi^0} v_x^0$$

$$P2 = D\varphi' f_6(\phi^0)$$

$$Q1 = -D\varphi' \frac{df_7(\phi^0)}{d\phi^0}(v_x^0)^2 + D(1-e_w^2)\frac{df_8(\phi^0)}{d\phi^0}T^0$$

$$Q3 = -D\varphi' f_7(\phi^0)(v_x^0)^2$$

$$Q4 = D(1-e_w^2) f_8(\phi^0) T^0$$

$$R1 = \frac{df_9(\phi^0)}{dy}(u_x^0 - v_x^0)\frac{BD(1-E)G}{C} + \frac{BD}{C\phi_{\max}}\frac{dP_g^0}{dx} - \frac{d\mu_{eg}(\phi^0)}{d\phi^0}\frac{du_x^0}{dy} - \frac{\phi_{\max}}{G}2T^0 u_x^0 \frac{df_{10}(\phi^0)}{dy} + \frac{BDEG}{C\phi_{\max}}$$

$$R2 = f_9(\phi^0)\frac{BD(1-E)G}{C} - \frac{\phi_{\max}}{G}2T^0 f_{10}(\phi^0)$$

$$R3 = -\mu_{eg}(\phi^0)$$

$$R4 = -f_9(\phi^0)\frac{BD(1-E)G}{C}$$

$$R5 = -\frac{\phi_{\max}}{G}2 f_{10}(\phi^0) u_x^0$$

$$R6 = \frac{BDG}{C}\frac{\phi_{\max}}{\phi} ik_x$$

Dimensionless Functions of $f_0 - f_{10}$

$$f_0 = \phi(1-\phi)^{1-N}$$

$$f_{02} = \frac{81\mu_{eg}^2}{g_0\sqrt{\pi}}\phi$$

$$f_1 = \phi(1+4\eta\phi g_0)$$

$$f_2 = \frac{16(2+\alpha)}{15}\eta\frac{\phi^2 g_0}{\sqrt{\pi}}$$

$$f_3 = \frac{2+\alpha}{3}[\frac{5\sqrt{\pi}}{48g_0\eta(2-\eta)}(1+\frac{8}{5}\eta\phi g_0)(1+\frac{8}{5}\eta(3\eta-2)\phi g_0)]$$

$$f_4 = \frac{25\sqrt{\pi}}{16\eta(41-33\eta)g_0}[(1+\frac{12}{5}\eta\phi g_0)(1+\frac{12}{5}\eta^2(4\eta-3)\phi g_0)+\frac{64}{25\pi}(41-33\eta)(\eta\phi g_0)^2]$$

$$f_5 = \frac{48}{\sqrt{\pi}}\eta(1-\eta)\phi^2 g_0$$

$$f_6 = \frac{\sqrt{3}}{3}\frac{\pi\phi g_0}{\phi_{\max} f_3}$$

$$f_7 = \frac{\sqrt{3}}{6}\frac{\pi\phi g_0}{\phi_{\max} f_4}$$

$$f_8 = \frac{\sqrt{3}}{4}\frac{\pi\phi g_0}{\phi_{\max} f_4}$$

$$f_9 = \frac{\phi}{\phi_{\max}}\phi(1-\phi)^{1-N}$$

$$f_{10} = \frac{\mu_{eg}}{\phi}$$

FSC
www.fsc.org
MIX
Papier | Fördert gute Waldnutzung
FSC® C083411

Zeitfracht Medien GmbH
Ferdinand-Jühlke-Straße 7
99095 Erfurt, Deutschland
produktsicherheit@kolibri360.de

Druck:
CPI Druckdienstleistungen GmbH
im Auftrag der
Zeitfracht Medien GmbH
Ein Unternehmen der Zeitfracht - Gruppe
Ferdinand-Jühlke-Str. 7
99095 Erfurt